SOLUTIONS MANUAL TO ACCOMPANY FINITE MATHEMATICS

SOLUTIONS MANUAL TO ACCOMPANY FINITE MATHEMATICS: MODELS AND APPLICATIONS

CARLA C. MORRIS
University of Delaware

ROBERT M. STARK
University of Delaware

WILEY

Published by John Wiley & Sons, Inc., Hoboken, New Jersey
Published simultaneously in Canada

Library of Congress Cataloging-in-Publication Data:

Morris, Carla C.
Finite Mathematics : Models and Applications / Carla C. Morris and Robert M. Stark.
pages cm
Includes bibliographical references and index.
ISBN 978-1-119-01550-5 (cloth)
1. Mathematics–Textbooks. I. Stark, Robert M., 1930- II. Title.
QA39.3.M68 2015
511′.1–dc23

2014042187

Printed in the United States of America

CONTENTS

CHAPTER 1

LINEAR EQUATIONS AND MATHEMATICAL CONCEPTS

EXERCISES 1.1

1. $3x + 1 = 4x - 5$

$1 = x - 5$

$x = 6$ conditional equation

3. $5(x + 1) + 2(x - 1) = 7x + 6$

$5x + 5 + 2x - 2 = 7x + 6$

$7x + 3 = 7x + 6$ contradiction

5. $4(x + 3) = 2(2x + 5)$

$4x + 12 = 4x + 10$ contradiction

Solutions Manual to Accompany Finite Mathematics: Models and Applications, First Edition.
Carla C. Morris and Robert M. Stark.

7. $5x - 3 = 17$

$5x = 20$

$x = 4$

9. $2x = 4x - 10$

$2x - 4x = -10$

$-2x = -10$

$x = 5$

11. $4x - 5 = 6x - 7$

$-5 + 7 = 6x - 4x$

$2 = 2x$

$1 = x$

13. $0.6x = 30$

$x = 30/0.60 = 50$

15. $2/3 = (4/5)x - (1/3)$ multiply by 15 to eliminate fractions

$1\,(2/3) = 15\,\{(4/5)x - (1/3)\}$

$10 = 12x - 5$

$15 = 12x$

$5/4 = x$

17. $5(x - 4) = 2x + 3(x - 7)$

$5x - 20 = 2x + 3x - 21$

$5x - 20 = 5x - 21$

No solution

19. $3s - 4 = 2s + 6$

$s - 4 = 6$

$s = 10$

21. $7t + 2 = 4t + 11$

$$7t - 4t = 11 - 2$$

$$3t = 9$$

$$t = 3$$

23. $4(x + 1) + 2(x - 3) = 7(x - 1)$

$$4x + 4 + 2x - 6 = 7x - 7$$

$$6x - 2 = 7x - 7$$

$$6x - 7x = -7 + 2$$

$$x = -5$$

$$x = 5$$

25. $\dfrac{x + 8}{2x - 5} = 2$ multiply by $2x - 5$ to eliminate the fraction

$$(x + 8) = 2(2x - 5)$$

$$x + 8 = 4x - 10$$

$$8 + 10 = 4x - x$$

$$18 = 3x$$

$$6 = x$$

(Check the result. Multiplication by a factor such as $2x - 5$ can introduce an extraneous solution.)

27. $8 - \{4[x - (3x - 4) - x] + 4\} = 38 - \{4[x - (3x - 4) - x] + 4\}$

$$= 3(x + 2)$$

$$8 - \{4[x - 3x + 4 - x] + 4\} = 3x + 6$$

$$8 - \{4[-3x + 4] + 4\} = 3x + 6$$

$$8 - \{-12x + 16 + 4\} = 3x + 6$$

$$8 - \{-12x + 20) = 3x + 6$$

$$8 + 12x - 20 = 3x + 6$$

$$12x - 12 = 3x + 6$$

$$9x = 18$$

$$x = 2$$

29. $6x - 3y = 9$ for x

$$6x = 3y + 9$$

$$x = \frac{3y+9}{6} = \frac{1}{2}y + \frac{3}{2}$$

31. $3x + 5y = 15$

$$5y = 15 - 3x$$

$$y = \frac{(15 - 3x)}{5}$$

$$y = 3 - \left(\frac{3}{5}\right)x$$

33. $V = LWH$

$$\frac{V}{LH} = W$$

35. $Z = \dfrac{(x - \mu)}{\sigma}$

$$Z\sigma = x - \mu$$

$$x = Z\sigma + \mu$$

37. Let x = monthly installment ($).
Since Sally paid $300 down, she owes $1300 – $300 = $1000.
Therefore, $5x = 1000$ or $x = \$200$ is the monthly installment.

39. The consumption function is $C(x) = mx + b$. The slope is the "marginal propensity to consume." Therefore, $C(x) = 0.75x + b$. The disposable income, $x = 2$, when consumption is $y = 11$ yields $11 = (0.75)(2) + b$ and $b = 9.5$. The consumption function is $C(x) = 0.75x + 9.5$.

41. a) $d = 4.5(2) = 9$ miles
b) $18 = 4.5t$ and $t = 18/4.5 = 4$ seconds

43. The tax is 6.2% or 0.062 in decimal form, so $T = 0.062x$, where x is $0 \le x \le 87{,}000$.

45. a) BSA = 1321 + (0.3433)(20,000) = 8187 cm^2
b) 10,325 = 1321 + (0.3433)(Wt)
9004 = (0.3433)(Wt)
9004/0.3433 = 26,228 g = 26.2 kg

EXERCISES 1.2

1. Setting $y = 0$ determines the x-intercept and setting $x = 0$ determines the y-intercept.

a) $5x - 3y = 15$	x-intercept 3, y-intercept -5
b) $y = 4x - 5$	x-intercept 5/4, y-intercept -5
c) $2x + 3y = 24$	x-intercept 12, y-intercept 8
d) $9x - y = 18$	x-intercept 2, y-intercept -18
e) $x = 4$	x-intercept 4, no y-intercept(vertical line)
f) $y = -2$ no	x-intercept (horizontal line), y-intercept -2

3. The slope is $m = \dfrac{y_2 - y_1}{x_2 - x_1}$

a) (3, 6) and (−1, 4) $\quad m = \dfrac{4-6}{-1-3} = \dfrac{-2}{-4} = \dfrac{1}{2}$

b) (1, 6) and (2, 11) $\quad m = \dfrac{11-6}{2-1} = \dfrac{5}{1} = 5$

c) (6, 3) and (12, 7) $\quad m = \dfrac{7-3}{12-6} = \dfrac{4}{6} = \dfrac{2}{3}$

d) (2, 3) and (2, 7) $\quad m = \dfrac{7-3}{2-2} = \dfrac{4}{0}$ undefined

e) (2, 6) and (5, 6) $\quad m = \dfrac{6-6}{5-2} = \dfrac{0}{3} = 0$

f) (5/3, 2/3) and (10/3, 1) $m = \dfrac{1 - {}^{2}\!/_{3}}{{}^{10}\!/_{3} - {}^{5}\!/_{3}} = \dfrac{{}^{1}\!/_{3}}{{}^{5}\!/_{3}} = \dfrac{1}{5}$

5. a) x-intercept 5/2 and y-intercept -5

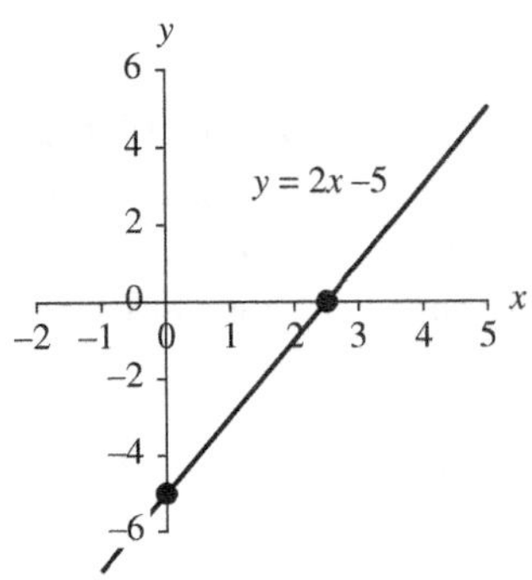

b) x-intercept 4 and no y-intercept

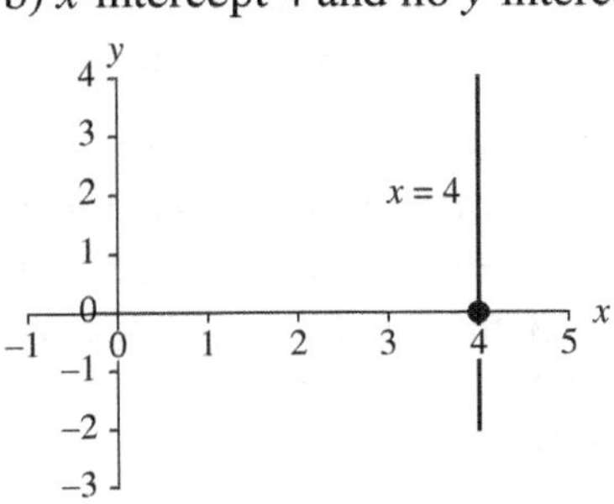

⟶

c) x-intercept 5 and y-intercept 3

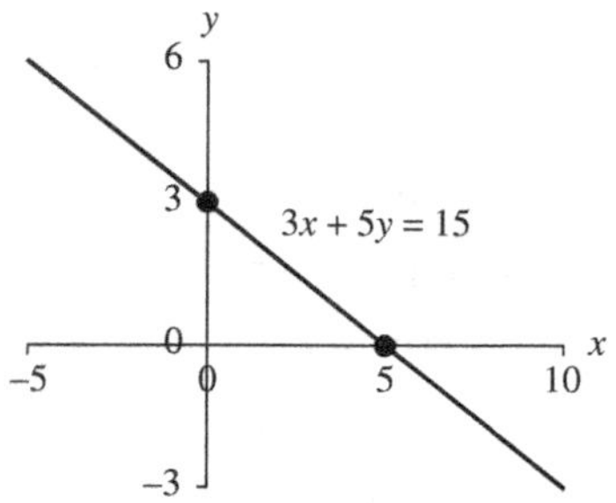

d) x-intercept 7 and y-intercept 2

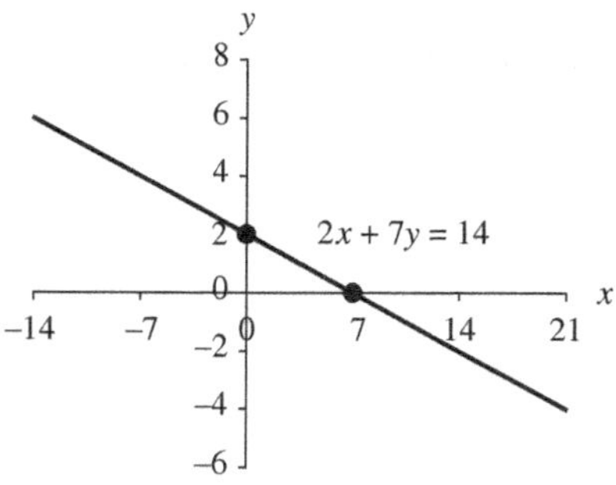

7. a) $y = (5/3)x + 2$ and $5x - 3y = 10$; the slope of the first line is 5/3. Solving for y in the second equation yields $y = (5/3)x - (10/3)$. This slope is also 5/3. The slopes are both (5/3) so the lines are parallel (with different intercepts).

b) $6x + 2y = 4$ and $y = (1/3)x + 1$. The slope of the second line is easily determined (line in slope intercept form) as 1/3. Again, solve for y in the first equation to determine $y = -3x + 2$. The slope is -3. The slopes are negative reciprocals; the lines are perpendicular.

c) $2x - 3y = 6$ and $4x - 6y = 15$. Solving for y in each equation, one determines that $y = (2/3)x - 2$ and $y = (2/3)x - (5/2)$. These lines have the same slope (and different intercepts) making them parallel.

d) $y = 5x - 4$ and $3x - y = 4$. The slope of the first line is 5 and solving for y in the second equation, ($y = 3x - 4$), the slope is 3. These slopes are neither the same nor negative reciprocals. They are neither parallel nor perpendicular.

e) $y = 5$ is a horizontal line while $x = 3$ is a vertical line. The two lines are perpendicular.

9. A linear equation has a single x-intercept except for $y = 0$ (the x-axis) with an infinite number of x-intercepts. Any horizontal line except $y = 0$ has no x-intercepts. Generally, lines do not have more than one y-intercept. The exception is $x = 0$ (the y-axis) with an infinite number of y-intercepts. Any vertical line with the exception of $x = 0$ has no y-intercepts.

11. The ordered pairs of "time" and "machine value" are (0, 75,000) and (9, 21,000), respectively. The slope is

$m = \dfrac{21{,}000 - 75{,}000}{9 - 0} = \dfrac{-54{,}000}{9} = -6000$. The y-intercept is the purchase price, \$75,000. The equation to model the straight-line depreciation is $V(t) = -6000t + 75{,}000$, where $V(t)$ is the machine value (\$) at time t.

13. The ordered pairs (gallons, miles) are (7, 245) and (12, 420).
The slope is $\dfrac{420 - 245}{12 - 7} = \dfrac{175}{5} = 35$ with x gallons and y miles.
Use either pair with the point slope-formula.
Therefore, $y - 245 = 35(x - 7)$ or $y = 35x$.

15. Total cost reflects both fixed and variable costs. The fixed cost is monthly rent (\$1100). The variable cost is $5x$, where x is monthly production. Therefore, total cost is $C(x) = 1100 + 5x$.

17. a) Here, the fixed cost is \$50/day and variable cost \$0.30/mile. To rent the car for a single day costs \$50 to which the mileage cost must be added. The cost is $C(x) = 50 + 0.30x$.

b) If a person has \$110 for rental, the equation to solve for the travel distance is $110 = 50 + 0.30x$.
Solving yields,

$$60 = 0.30x$$
$$\frac{60}{0.30} = x$$
$$200 = x$$

The person can rent the car and travel 200 miles with \$110.

19. Since R is to be a function of C, the ordered pairs are (C, R). The two ordered pairs are (70, 84) and (40, 48). The slope is $\dfrac{48 - 84}{40 - 70} = \dfrac{36}{30} = \dfrac{6}{5}$. Using either pair with the slope to yield $R - 84 = (6/5)(C - 70)$ or $R = (6/5)C$.

EXERCISES 1.3

1. The ordered pair must satisfy each equation to be a solution to the system.

a) $2(3) + 1 = 7$ is true but $3 + 1 = 5$ is not. Therefore, (3, 1) is not a solution to the system.

b) $2(2) + 3 = 7$ is true and so is $2 + 3 = 5$. Therefore, (2, 3) is a solution to the system.

c) $2(4) + 1 = 7$ is true but $4 + (-1) = 5$ is not. Therefore, (4, −1) is not a solution to the system.

3. a) $y = (-1/3)x + (8/3)$ and $y = -x + 6$, $m_1 = -1/3$ and $m_2 = -1$. Since the slopes differ, this is a consistent system.

b) $y = (-1/2)x + (7/2)$ and $y = (-1/2)x + (7/2)$. Since both the slopes and intercepts are the same, the two equations are the same line. It is a dependent system.

c) $y = (-3/2)x + (7/2)$ and $y = (-3/2)x + 5$. Here, the slopes are the same and the intercepts differ. The lines are parallel and the system is inconsistent.

5. The graphs and solution to each system are:

a)

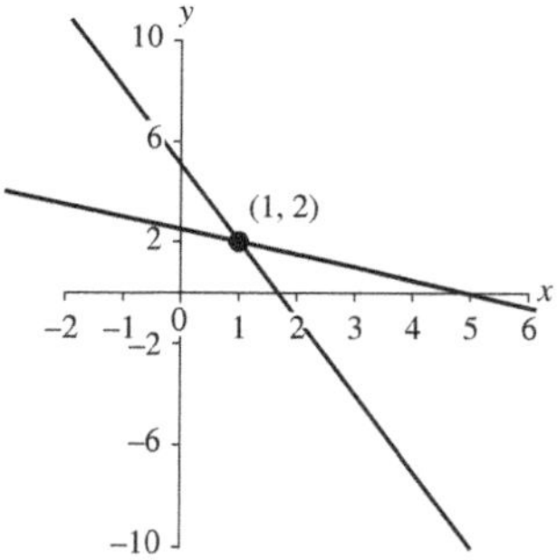

b)

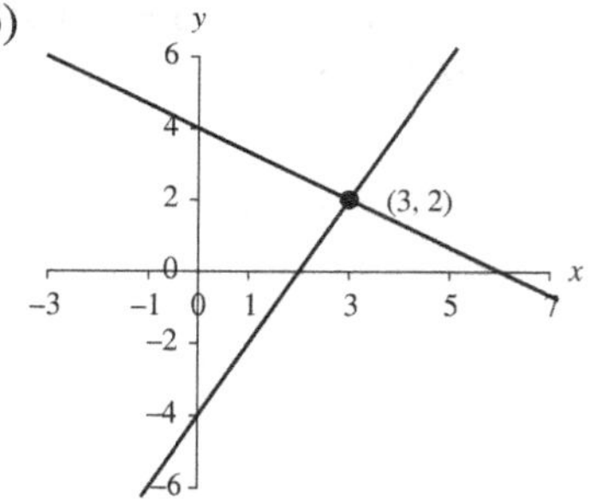

c)

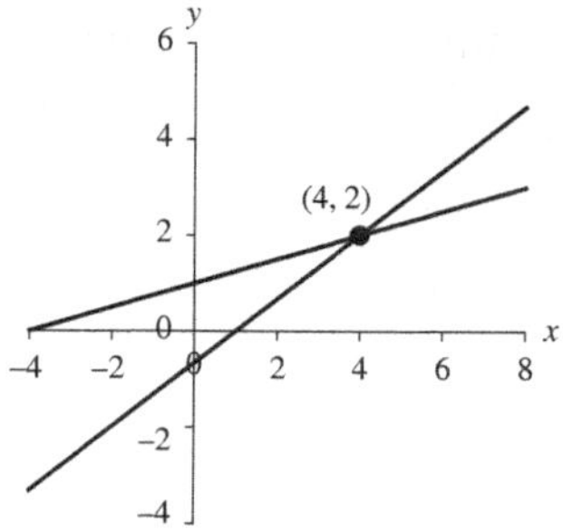

7. a) Given $x = 3$ and substituting in the second equation yields $3 + 3y = 9$, $3y = 6$, and $y = 2$. The ordered pair solution is (3, 2).

b) Using, $y - 2 = 0$, $y = 2$. Substituting into the second equation yields $x + 3(2) = 9$, $x + 6 = 9$, and $x = 3$. The ordered pair solution is (3, 2).

c) The second equation, already solved for y, substituted in the first equation yields $x + (-x + 3) = 5$ or $3 = 5$. This is false, so the system is inconsistent and has no solution.

9. a) Here, x can be eliminated by simply adding the two equations as written.

$$\begin{array}{r} -x + 2y = 5 \\ \underline{x + \ y = 4} \\ 3y = 9 \end{array} \text{ or } y = 3$$

Next, use $y = 3$ to determine $x = 1$. The ordered pair solution is (1, 3).

b) Here, y can be eliminated by multiplying the second equation by 3. The system is rewritten as

$$\begin{aligned} 4x + 3y &= 35 \\ 6x - 3y &= 15 \end{aligned}$$

Adding the two equations yields $10x = 50$ or $x = 5$. Using this value for x yields that y is also 5. (Check the solution in the original equation to see that (5, 5) is correct.)

c) The second equation must first be rewritten in standard form $ax + by = c$, so the system to solve is

$$\begin{aligned} x + 4y &= 13 \\ 4x + 2y &= 10 \end{aligned}$$

Multiplying the first equation by −4 will allow x to be eliminated from the system.

$$\begin{array}{r} -4x - 16y = -52 \\ \underline{4x + \ \ 2y = 10} \\ -14y = -42 \end{array} \text{ or } y = 3$$

Using $y = 3$, it is determined that $x = 1$. (Check the solution in the original system to verify the solution (1, 3)).

⟶

d) Rewriting the first equation in standard form and multiplying by 2 to eliminate x yields

$$\begin{array}{r} -2x + 2y = 8 \\ 2x + 3y = 12 \\ \hline 5y = 20 \text{ or } y = 4 \end{array}$$

If $y = 4$, then $x = 0$ and checking in the original system indicates that (0, 4) is the correct solution.

11. Let x represent the number of boxes of cookies and y the boxes of candy. The system to be solved is

$$x + y = 2400$$

$$4x + 5y = 10{,}500$$

Solving yields sales of 1500 boxes of cookies and 900 boxes of candy.

13. Let x = gallons of regular and y = gallons premium. The system of equations to be solved (by either substitution or elimination) is

$$x + y = 100$$

$$0.87x + 0.93\,y = 0.92\ (100)$$

Using elimination,

$$\begin{array}{r} -0.87x - 0.87y = -87 \\ 0.87x + 0.93y = 92 \\ \hline 0.06y = 5 \end{array}$$

$y = 83\ 1/3$ gallons and $x = 16\ 2/3$ gallons

15.

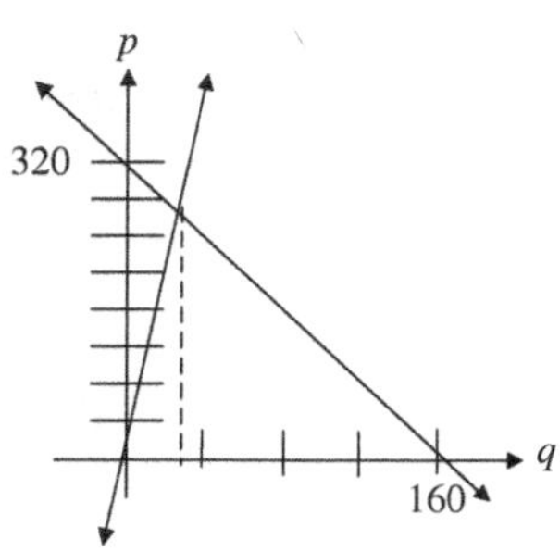

The Market equilibrium occurs when q is about 30 and y about 260. Using substitution

$$-2q + 320 = 8q + 20$$
$$300 = 10q$$
$$30 = q$$
$$\text{so, } p = 260$$

17.

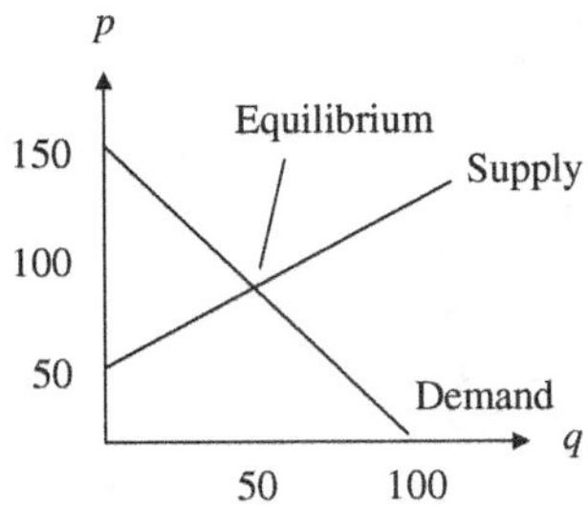

The price is about \$75 and quantity is 50 at market equilibrium.

EXERCISES 1.4

1. The area of a circle is πr^2, so,

$$\pi r^2 = \pi$$
$$r^2 = 1$$
$$r = 1$$

The radius is 1 unit, so the diameter is two units.

3. a) Pi day is March 14 (3/14)

b) Professor Yasumasa Kanada of the University of Tokyo.

c) Albert Einstein was born on March 14.

d) It is an illness thought to be attached to trying to square a circle.

e) Pi is known as die Ludolphschezahl in Germany.

5. a) $\left(1+\frac{1}{5}\right)^5 = 2.48832$ b) $\left(1+\frac{1}{10}\right)^{10} = 2.59374$

c) $\left(1+\frac{1}{25}\right)^{25} = 2.66584$

EXERCISES 1.5

1. a) The expression must involve powers of 2 or 3. In this case, 8 is a power of 2, so one has $(8)^{3x} = (2^3)^{3x} = 2^{9x}$.

b) Rewrite as $(27)^{2x} = (3^3)^{2x} = 3^{6x}$.

c) Rewrite as $(16)^{5x} = (2^4)^{5x} = 2^{20x}$.

3. a) $\left(\frac{1}{8}\right)^{-4x} = \left(\frac{1}{2^3}\right)^{-4x} = (2^{-3})^{-4x} = (2)^{12x}$

b) $\left(\frac{1}{9}\right)^{6x} = \left(\frac{1}{3^2}\right)^{6x} = (3^{-2})^{6x} = (3)^{-12x}$

c) $\left(\frac{1}{27}\right)^{-2x} = \left(\frac{1}{3^3}\right)^{-2x} = (3^{-3})^{-2x} = (3)^{6x}$

5. a) $\frac{10^{5x}}{5^{5x}} = (2)^{5x}$

b) $\frac{32^{2x}}{16^{2x}} = (2)^{2x}$

c) $\frac{4^{3x}}{12^{3x}} = \left(\frac{1}{3}\right)^{3x} = (3^{-1})^{3x} = (3)^{-3x}$

7. $\frac{7x^3x^5y^{-2}}{x^2y^4} = \frac{7x^8}{x^2y^4y^2} = \frac{7x^6}{y^6}$

9. $\frac{x^3}{y^{-2}} \div \frac{x}{y^5} = (x^3y^2) \cdot \left(\frac{y^5}{x}\right) = x^2y^7$

11. $\frac{2^{5x+3}4^{x+1}}{8(2^{3x-1})} = \frac{2^{5x+3}(2^2)^{x+1}}{2^3(2^{3x-1})} = \frac{2^{5x+3}2^{2x+2}}{2^{3x+2}} = 2^{4x+3}$

13. Since the bases are the same, exponents are equal. Therefore, $3x = 15$ yields $x = 5$.

15. Equate bases as $2^{7-x} = 2^5$ to yield $7 - x = 5$ and $x = 2$.

17. Factoring yields $5^x[(1 + x) + (3 - 2x)] = 0$

$$5^x[(4 - x)] = 0$$

Since 5^x cannot be zero, $4 - x = 0$ and $x = 4$.

19. Factoring yields $7^x[(x^2 + 4x) + (x + 6)] = 0$. Simplifying yields $7^x[x^2 + 5x + 6] = 0$. Therefore, $[x^2 + 5x + 6] = (x + 3)(x + 2) = 0$. The solutions are $x = -2$ or $x = -3$.

21. $2^{3+h} = 2^h(2^3)$

23. $7^{x+5} - 7^{2x} = 7^{2x}(7^{x+5} - 1)$

25. This is a difference of cubes $(7^h)^3 - 8 = (7^h - 2)[7^{2h} + 2(7^h) + 4]$

27. Rewrite as $e^y = x$ to determine ordered pairs and graph as

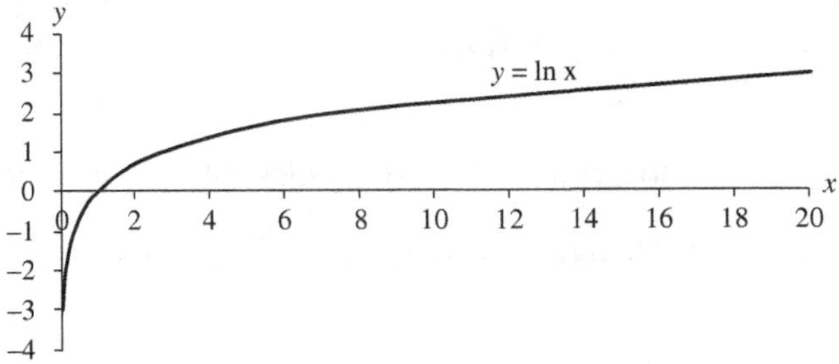

29. $\log_3 243 = x$ so $3^x = 3^5$, which yields $x = 5$.

31. $\log_5 125 = x$ so $5^x = 125$, which yields $x = 3$.

33. $\log_3 \dfrac{1}{81} = x$ so $3^x = 3^{-4}$, which yields $x = -4$

35. Given $\ln e^x = x$, then $\ln e^7 = 7$.

37. Given $\ln e^x = x$, then $\ln e^{-3.4} = -3.4$.

39. $e^{\ln 1} = e^0 = 1$

41. $\log_{25} 125 = x$ so $25^x = 125$, which yields $5^{2x} = 5^3$.
Therefore, $2x = 3$ or $x = 3/2$.

43. $\log_5 625 = x$ so $5^x = 625 = 5^4$. Therefore, $x = 4$.

45. $\log_4(1/64) = x$ so $4^x = 4^{-3}$. Therefore, $x = -3$.

47. Rewriting $\log_x 64 = 3$ in exponential form yields $x^3 = 64$ and $x = 4$.

49. Rewriting $\log_2(x^2 + 7x) = 3$ yields $2^3 = x^2 + 7x$. Solving $x^2 + 7x - 8 = 0$ yields $(x + 8)(x - 1) = 0$ and $x = 1$ or $x = -8$.

51. Rewrite the equation as $e^{5x} = 2$. Taking natural logarithms of both sides yields $5x = \ln 2$ or $x = (1/5)\ln 2$.

53. In exponential form $e^{1/3} = 7 - x$. Solving for x yields $x = 7 - e^{1/3}$.

55.
$$\ln \frac{(x-1)^4}{(2x+3)^5(x-4)^2} = \ln (x-1)^4 - \ln (2x+3)^5 - \ln (x-4)^2$$
$$= 4\ln(x-1) - 5\ln(2x+3) - 2\ln(x-4)$$

57. Using the rules for logarithms yields $\ln \dfrac{2(7)}{3} = \ln \dfrac{14}{3}$.

59. Using logarithms, $10^{3x-1} = 100{,}100$ is written as $\log_{10}10^{3x-1} = \log_{10}100{,}100$. This yields $3x - 1 = \log_{10}100{,}100$. Therefore, $x = \dfrac{1 + \log_{10}100{,}100}{3}$.

61. Taking natural logarithms on both sides of the equation yields $\ln e^{3x+1} = \ln 22$. Therefore, $3x + 1 = \ln 22$ and $x = \dfrac{-1 + \ln 22}{3}$.

EXERCISES 1.6

1. a) Direct variation. The greater the speed, the farther the distance traveled.

b) Inverse variation. The greater the speed, the less time to complete the race.

c) Direct variation. The larger the income, the higher the tax rate.

d) Direct variation. The larger the area of a circle, the larger its diameter.

e) Direct variation. The larger the distance on the blue print, the larger the actual distance.

3. Here, $y = kr^3$. Use this to determine k.
Therefore, $32 = k(4)^3$ yields 1/2 for k. When $r = 6$,
$y = (1/2)(6)^3 = (1/2)(216) = 108$.

5. Here, $y = kx$. Use this to determine k.
Therefore, $15 = k(10)$ yields 3/2 for k. When $y = 22.5$,
$x = (22.5)/(3/2) = (45/2)(2/3) = 15$.

7. Since x varies directly as y, let $x = ky$. Values appear in the table.

x	2	4	8
y	12.5	25	50

Therefore, $2 = k(12.5)$ yields $k = 0.16$. The direct relationship is $x = (0.16)y$. As a check, $4 = 0.16(25)$ and $8 = (0.16)50$ are true.

9. The area of a circle with diameter d is $A = kd^2$. Here,
$36\pi = k(12)^2 = 144k$ and $k = \pi/4$. The area of a circle is $A = \pi d^2/4$.

11. Here, $V = \frac{1}{3}\pi r^2 h$. Using the data
$V = \frac{1}{3}\pi(3)^2(10) = 30\pi\,\text{cm}^3$.

13. Here, Hooke's Law $y = kx$, $400 = k(10)$ and $k = 40$. When the force is 250 Newtons, $250 = (40)x$ and $x = 250/40 = 6.25$ cm.

15. Let x represent the length of the shorter leg and y the length of the hypotenuse. To determine x, use $5/x = 12/24$ and to determine y, use $13/y = 12/24$. Solving yields

$$\frac{5}{x} = \frac{12}{24} \qquad \frac{13}{y} = \frac{12}{24}$$

$$12x = 120 \qquad 12y = 312$$

$$x = 10 \qquad y = 26$$

17. Here, the proportionality yields the equation

$$\frac{x}{x+5} = \frac{20}{60}$$

$$\frac{x}{x+5} = \frac{1}{3}$$

$$3x = x + 5$$

$$2x = 5$$

$$x = 2.5 \quad \text{and} \quad x + 5 = 7.5$$

19. Here, $A = kbh$ so, $50 = k(20)(5)$ to yield $k = 1/2$. The area of the triangle of interest $A = (1/2)(16)(8) = 64\,\text{in.}^2$.

21. NASA CONNECT is an annual series of math, science, and technical instructional distance programs for students in grades 6–8. In proportionality modeling of the future in March 2000 students learned why scaling and proportion are important in the design of aircraft transportation systems.

23. Answers vary.

EXERCISES 1.7

1. $5\,\text{miles} \cdot \dfrac{5280\,\text{ft}}{1\,\text{mile}} \cdot \dfrac{12\text{in.}}{1\,\text{ft}} \cdot \dfrac{1\,\text{m}}{39.37\,\text{in}} = 8046.74\,\text{m}$

3. Here, use the conversion factor from Exercise 1.4.1 (1 m = 39.37 in.) to solve.

$$\frac{100\ \text{km}}{1\ \text{h}} \cdot \frac{1000\ \text{m}}{1\ \text{km}} \cdot \frac{39.37\ \text{in.}}{1\ \text{m}} \cdot \frac{1\ \text{ft}}{12\ \text{in.}} \cdot \frac{1\ \text{mile}}{5280\ \text{ft}} = \frac{62.14\ \text{miles}}{\text{h}}$$

5. $\dfrac{50\ \text{miles}}{1\ \text{h}} \cdot \dfrac{1\ \text{h}}{60\ \text{min}} \cdot \dfrac{1\ \text{min}}{60\ \text{sec}} \cdot \dfrac{5280\ \text{ft}}{1\ \text{mile}} \cdot \dfrac{12\ \text{in}}{1\ \text{ft}} \cdot \dfrac{1\ \text{m}}{39.37\ \text{in.}} \cdot \dfrac{100\ \text{cm}}{1\ \text{m}}$
$\approx 2235.2\ \text{cm/sec}$

7. $4.3\ \text{pounds} \cdot \dfrac{453.6\ \text{g}}{1\ \text{pound}} = 1950.48\ \text{g}$

9. $\dfrac{320.85\ \text{g}}{45\ \text{ml}} = 7.13\ \text{g/ml}$

11. The mass is the density times the volume (one liter here)
$\dfrac{13.55\,\text{g}}{\text{cm}^3} \times 1000\ \text{ml} = 13{,}550\ \text{g}$. Next, grams are converted into pounds
$13{,}550\ \text{g} \times \dfrac{1\ \text{lb}}{453.6\ \text{g}} = 29.872\ \text{lb}$

13. $30\text{gC} \cdot \dfrac{1\ \text{molC}}{12\,\text{gC}} \cdot \dfrac{6.02 \times 10^{23}\,\text{atomsC}}{1\,\text{molC}} = 1.505 \times 10^{24}\,\text{atoms C.}$

15. First, convert 5 tons to 10,000 pounds. Next, since volume is mass/density, $\dfrac{10{,}000\ \text{pounds}}{5.83\ \text{pounds/gallon}} = 1715.266\ \text{gallons.}$

17. Assuming 100 g of nitrous oxide, there is 63.6 g N and 36.4 g O. Converting these to moles, there is approximately 4.543 moles N and 2.275 moles of O implying that the formula is two parts N to one part O or N_2O. This formula weight is 44 g, which is the correct weight, so the subscripts are correct and nitrous oxide's chemical formula is N_2O.

19. Acres are measured by square feet. An acre can have various dimensions with the area approximately 43,560 square feet.

SUPPLEMENTARY EXERCISES CHAPTER 1

1. To determine an average, the five scores are added and divided by five. If x represents the score on the final exam, then

$$\frac{76 + 89 + 70 + x + x}{5} = 82.6$$

$$235 + 2x = 413$$

$$2x = 178$$

$$x = 89$$

The student will need to score an 89 to receive a B in the course.

3. a) A single filer with \$250,000 net taxable income has a tax of 10% on the first \$9075, 15% for next \$27,825 and so on until the \$250,000 is met.(9075)(0.10) + 27,825(0.15) + 52,450(0.25) + 97,000(0.28) + 63,650(0.33) = \$66,358 tax.

b) A married couple filing jointly with a \$250,000 taxable income has a tax of 18,150(0.10) + 55,650(0.15) + 75,050(0.25) + 78,000(0.28) + 23,150(0.33) = \$58,405.

c) A married person filing separately with \$200,000 taxable income would pay \$53,952 and the spouse making \$50,000 would pay \$8356 for a total tax of \$62,308.
It would not be advantageous for the couple to file separately in this case.

5. Two points on the line are (0, 8) and (5, 0). The slope is $-8/5$ since

$$m = \frac{0 - 8}{5 - 0} = -\frac{8}{5}$$

The intercept b is 8, so $y = (-8/5)x + 8$ is the equation for the line shown.

7. a) Perpendicular lines have negative reciprocal slopes. The slope of the line through (2, 0) and (0, 5) is $m = \dfrac{5 - 0}{0 - 2} = -5/2$ and the slope of a perpendicular line is $m = 2/5$. The equation of a line with slope 2/5 that passes through (3, 5) using the point-slope form, is $y - 5 = (2/5)(x - 3)$.

b) A horizontal line has the form $y = b$. Here, the y value must be 4, since $(1, 4)$ is to be on the line. The equation is $y = 4$.

⟶

c) A line perpendicular to the x-axis is a vertical line. Since (2, 7) is on the line, the equation is $x = 2$.

9. The line has a negative slope, since it "falls". It also has a negative y-intercept. The line is d) $-2x - 3$.

11. Let x = number of gallons of water used (in thousands), then the bill is $40 + 2.25x$. Here, use $x = 12$ to yield $40 + 2.25(12) = \$67$.

13. Let x = amount of water and y = amount of 100% sulfuric acid. The system of equations to solve is

$$x + y = 50$$

$$0x + 100y = 20\ (50)$$

The second equation yields $y = 10$ so $x = 40$.
The mix must be 40 ml of water and 10 ml of pure acid.

15. Let x = # \$50 items and y = # \$100 items. The system to solve is

$$x + y = 25$$

$$50x + 100y = 1650$$

Using elimination,

$$\begin{array}{r} -50x - 50y = -1250 \\ \underline{50x + 100y = 1650} \\ 50y = 400 \text{ so } y = 8 \text{ and } x = 17. \end{array}$$

Nine more \$50 items were sold than \$100 items. (Take care to answer the question.)

17. a) $R(x) = 6x$ b) $C(x) = 2000 + 2x$

c) Setting revenue = cost for the breakeven point, $6x = 200 + 2x$, so $x = 500$.

d) $P(x) = 6x - (2000 + 2x) = 4x - 2000$ and $P(1000) = 2000$ profit

19. a) The information is on the Internet .

b) There does not seem to be a digit that appears more frequently.

21. $\dfrac{125^{2x-3}25^{x+1}}{5^{4x-5}} = \dfrac{(5^3)^{2x-3}(5^2)^{x+1}}{5^{4x-5}} = \dfrac{5^{6x-9}5^{2x+2}}{5^{4x-5}} = 5^{4x-2}.$

23. Rewrite as $\log_4 128 = x$ so $4^x = 128$ or $2^{2x} = 2^7$.
Therefore, $2x = 7$ or $x = 7/2$.

25. First, simplify as $e^{x-1} = 5$. Next, logarithms of both sides yield $\ln(e^{x-1}) = \ln 5$. Therefore, $x - 1 = \ln 5$ or $x = 1 + \ln 5$.

27. Here, xy = constant and the first pair in the table indicates that the constant is (2)(24) = 48.

x	2	4	6	12	1
y	24	12	8	4	48

29. $$54\ \text{g C} \cdot \frac{1\ \text{mole C}}{12\ \text{g C}} \cdot \frac{6.02 \times 10^{23}\ \text{atoms C}}{1\ \text{mole C}} = 2.709 \times 10^{24}\ \text{atoms C}.$$

CHAPTER 2

MATHEMATICS OF FINANCE

EXERCISES 2.1

1. Here, the principal is \$5000, the rate is 0.08, and the time is 3 years. Therefore, $I = (5000)(0.08)(3) = \$1200$.

3. Here, the principal is \$50,000, the rate is 0.05, and the time is 7 years. Therefore, $I = (50{,}000)(0.05)(7) = \$17{,}500$.

5. Here, the principal is \$50,000, the rate is 0.04, and the time is 6 years. Therefore, $P_t = 50{,}000(1 + 0.04)^6 = \$63{,}265.95$.
The interest is found by subtracting the principal to yield \$13,265.95.

7. Here, the principal is \$500,000, the rate is 0.03, and the time is 6 years. Therefore, $P_t = 500{,}000(1 + 0.03)^6 = \$597{,}026.15$.
The interest is found by subtracting the principal to yield \$97,026.15.

9. Here, the principal is \$250,000, the rate is 0.05, and the time is 4 years. Therefore, $P_t = 250{,}000(1 + 0.05)^4 = \$303{,}876.56$.
The interest is found by subtracting the principal to yield \$53,876.56.

Solutions Manual to Accompany Finite Mathematics: Models and Applications, First Edition.
Carla C. Morris and Robert M. Stark.

11. Here, the principal is \$100,000, the rate is 0.06, n is 4, and the time, 10 years. Therefore,

$$P_t = 100{,}000\left(1 + \frac{0.06}{4}\right)^{4\cdot 10} = 100{,}000(1.015)^{40} = \$181{,}401.84.$$

13. Here, the principal is \$50,000, the rate is 0.03, n is 12, and the time, 7 years. Therefore,

$$P_t = 50{,}000\left(1 + \frac{0.03}{12}\right)^{12\cdot 7} = 50{,}000(1.0025)^{84} = \$61{,}667.74.$$

15. Here, the principal is \$75,000, the rate is 0.05, and the time, 5 years. Therefore, $P_t = 75{,}000e^{(0.05)(5)} = 75{,}000e^{0.25} = \$96{,}301.91$.

17. Here, the principal is \$75,000, the rate is 0.07, and the time, 6 years. Therefore, $P_t = 75{,}000e^{(0.07)(6)} = 75{,}000e^{0.42} = \$114{,}147.12$.

19. Here, the principal is \$50,000, the rate is 0.08, and the time, 3 years. Therefore, $P_t = 50{,}000e^{(0.08)(3)} = 50{,}000e^{0.24} = \$63{,}562.46$.
The interest is found by subtracting the principal to yield \$13,562.46.

21. Here, the principal is \$5000, the rate is 0.06, n is 12, and the time, 9 years.

Therefore, $P_t = 5000\left(1 + \frac{0.06}{12}\right)^{12\cdot 9} = 5000(1.005)^{108} = \$8568.50.$
The interest is \$3568.50.
Solving $3568.50 = 5000(r)(9)$ yields $r = 0.0793$.

23. Here, $P_t = P_0\left(1 + \frac{0.10}{4}\right)^{4\cdot 8} = P_0(1.025)^{32} = 2.2038P_0$. The interest is $1.2038P_0$. The equation is $1.2038P_0 = P_0(r)(8)$ and $r = 1.2038/8 = 0.1505$. The effective interest rate is 15.05%.

25. $P_0(1 + 0.09)^8 = 250{,}000$. Therefore, $P_0 = \dfrac{250{,}000}{(1.09)^8} = \$125{,}466.57$.

27. $P_0(1 + 0.05)^{10} = 750{,}000$. Therefore, $P_0 = \dfrac{750{,}000}{(1.05)^{10}} = \$460{,}434.94$.

EXERCISES 2.2

1. Here, $P = \$1000$, $n = 20$, and $i = 0.05$. The future value is

$$\text{FV} = \frac{(1 + 0.05)^{20} - 1}{0.05}(1000) = \frac{(1.05)^{20} - 1}{0.05}(1000) = \$33{,}065.95.$$

3. Here, $P = \$3000$, $n = 20$, and $i = 0.05$. The future value is

$$FV = \frac{(1+0.05)^{20}-1}{0.05}(3000) = \frac{(1.05)^{20}-1}{0.05}(3000) = \$99{,}197.86.$$

5. Here, $P = \$2500$, $n = 8$, and $i = 0.03$. The future value is

$$FV = \frac{(1+0.03)^{8}-1}{0.03}(2500) = \frac{(1.03)^{8}-1}{0.03}(2500) = \$22{,}230.84.$$

7. Here, $P = \$1000$, $n = 15$, and $i = 0.07$. The future value is

$$FV = \frac{(1+0.07)^{15}-1}{0.07}(1000) = \frac{(1.07)^{15}-1}{0.07}(1000) = \$25{,}129.02.$$

9. Here, $P = \$1000$, $n = (2)(25) = 50$, and $i = 0.10/2 = 0.05$.
The future value is

$$FV = \frac{(1+0.05)^{50}-1}{0.05}(1000) = \frac{(1.05)^{50}-1}{0.05}(1000) = \$209{,}348.00.$$

11. Here, $P = \$50$, $n = (12)(5) = 60$, and $i = 0.12/12 = 0.01$.
The present value is

$$PV = \frac{1-(1+0.01)^{-60}}{0.01}(250) = \frac{1-(1.01)^{-60}}{0.01}(50) = \$2247.75.$$

13. Here, $P = \$125$, $n = (4)(10) = 40$, and $i = 0.06/4 = 0.015$.
The present value is

$$PV = \frac{1-(1+0.015)^{-40}}{0.015}(125) = \frac{1-(1.015)^{-40}}{0.015}(125) = \$3739.48.$$

15. Here, $P = \$2500$, $n = (2)(8) = 16$, and $i = 0.08/2 = 0.04$.
The present value is

$$PV = \frac{1-(1+0.04)^{-16}}{0.04}(2500) = \frac{1-(1.04)^{-16}}{0.04}(2500) = \$29{,}130.74.$$

17. The present value of a \$25,000 monthly payment at 6 % for 20 years is

$$PV = \frac{1-(1+0.005)^{-240}}{0.005}(25{,}000) = \frac{1-(1.005)^{-240}}{0.005}(25{,}000)$$
$$= \$3{,}489{,}519.30.$$

Since this exceeds the \$2,500,000 lump sum payment, the 20-year monthly payments is better.

EXERCISES 2.3

1. Using the amortization formula, the payment is

$$\frac{(0.01)(500)}{\left(1-\frac{1}{(1+0.01)^6}\right)} = \frac{5}{1-(1.01)^{-6}} = \$86.27.$$

Period	Beginning Balance	Interest	Amortized Payment (Less Interest)	Payment
1	\$500.00	\$5.00	\$81.27	\$86.27
2	\$418.73	\$4.19	\$82.08	\$86.27
3	\$336.65	\$3.37	\$82.90	\$86.27
4	\$253.75	\$2.54	\$83.73	\$86.27
5	\$170.02	\$1.70	\$84.57	\$86.27
6	\$85.45	\$0.82*	\$85.45*	\$86.27
Total		\$17.62	\$500.00	\$517.62

*denotes "adjusted amount."

3. Here, $i = 0.06/12 = 0.005$. The monthly payment is

$$\frac{25{,}000(0.005)}{1-\frac{1}{(1.005)^{60}}} = \$483.32.$$

5. Here, $i = 0.04/12 = 0.00333$. The monthly payment is

$$\frac{35{,}000(0.00333)}{1-\frac{1}{(1.00333)^{72}}} = \$547.58.$$

7. Here, $i = 0.03/12 = 0.0025$. The monthly payment for 4 years is $\dfrac{18{,}000(0.0025)}{1 - \dfrac{1}{(1.0025)^{48}}} = \$398.42$. The total of payments is (48)(\$398.42) or \$19,124.16 for a 4-year loan. In the case of a 5-year loan, the monthly payments are $\dfrac{18{,}000(0.0025)}{1 - \dfrac{1}{(1.0025)^{60}}} = \$323.44$ for a total of (60)(323.44) or \$19,406.40. Therefore, the 4-year loan can save you $19{,}406.40 - 19{,}124.16 = \$282.24$.

9. Here, $i = 0.04/12 = 0.00333$. The monthly payment for 30 years is $\dfrac{350{,}000(0.00333)}{1 - \dfrac{1}{(1.00333)^{360}}} = \$1670.95$. For a 25-year mortgage, the monthly payments are $\dfrac{250{,}000(0.00333)}{1 - \dfrac{1}{(1.00333)^{300}}} = \$1847.43$.

11. There are two options for the town to consider:
Option 1 \$900,000 amortized over 50 years at 5% + \$3000 per year
Option 2 \$800,000 amortized over 50 years at 5% + \$7000 per year

Option 1

$$Y = \frac{(0.05)(900{,}000)}{1 - \dfrac{1}{(1.05)^{50}}} = \$49{,}299.06.$$

Annual cost option 1 is \$49,299.06 + \$3000 = \$52,299.06.

Option 2

$$Y = \frac{(0.05)(800{,}000)}{1 - \dfrac{1}{(1.05)^{50}}} = \$43{,}821.39.$$

Annual cost option 2 is \$43,821.39 + \$7000 = \$50,821.39.
The town should choose option 2 as it has a lower annual cost.

13. Answers vary.

EXERCISES 2.4

1. It is an arithmetic sequence with a common difference of 5. The next four terms are determined by adding 5 to the previous term to yield 17, 22, 27, and 32.

3. Here, the common difference is 15. The next five terms are 65, 80, 95, 110, and 125.

5. The first seven terms are 3, 9, 15, 21, 27, 33, and 39. Adding terms yields 147. Another way uses $S_n = \dfrac{n(a_1 + a_n)}{2}$, where $a_1 = 3$ and $a_n = 39$. Therefore,

$$S_7 = \frac{7(3+39)}{2} = \frac{7(42)}{2} = 147.$$

7. For the first five terms of the sequence whose n^{th} term is $a_n = \dfrac{3n+1}{2n+3}$, substitute $n = 1, n = 2, n = 3, n = 4$, and $n = 5$ to yield

$$a_1 = \frac{3(1)+1}{2(1)+3} = \frac{4}{5}, \quad a_2 = \frac{3(2)+1}{2(2)+3} = \frac{7}{7} = 1, \quad a_3 = \frac{3(3)+1}{2(3)+3} = \frac{10}{9},$$

$$a_4 = \frac{3(4)+1}{2(4)+3} = \frac{13}{11}, \text{ and } a_5 = \frac{3(5)+1}{2(5)+3} = \frac{16}{13}.$$

9. Successive terms are determined by multiplying by the common ratio 3. So, the next three terms are 54, 162, and 486.

11. Here, the common ratio is 1/3. The next three terms are 27, 9, and 3.

13. The first five terms are 4, 16, 64, 256, and 1024. The sum is 1364. Using $S_n = \dfrac{a_1(1-r^n)}{1-r}$ with $a_1 = 4$ and $r = 4$ yields

$$S_5 = \frac{4(1-4^5)}{1-4} = \frac{4(-1023)}{-3} = 1364.$$

15. The heights of the ball follow a geometric sequence. It reaches 100 ft after the first bounce, 50 ft after the second, and 25 ft after the third or $\dfrac{1}{2^3}(200)$.

17. It will accumulate to $5000(1.06)^4 = \$6312.38$. Therefore, it will earn \$1312.38 in interest.

19. The products of the first three successive pairs of Fibonacci numbers are $F_1F_2 + F_2F_3 + F_3F_4$. Next, using $F_1 = F_2$ and that $F_4 = F_2 + F_3$ rewrite the expression as

⟶

$(F_2)F_2 + F_2F_3 + F_3(F_2 + F_3)$. Now,
$(F_2)F_2 + F_2F_3 + F_3(F_2 + F_3) = (F_2)^2 + 2F_2F_3 + (F_3)^2 = (F_2 + F_3)^2 = (F_4)^2$.
Next, use the first five pairs of Fibonacci numbers
$F_1F_2 + F_2F_3 + F_3F_4 + F_4F_5 + F_5F_6$ and the previous result.

$$\begin{aligned}(F_1F_2 + F_2F_3 + F_3F_4) + F_4F_5 + F_5F_6 &= (F_4)^2 + F_4F_5 + F_5F_6 \\ &= (F_4)^2 + F_4F_5 + F_5(F_4 + F_5) \\ &= (F_4)^2 + 2F_4F_5 + (F_5)^2 \\ &= (F_4 + F_5)^2 = (F_6)^2.\end{aligned}$$

Continuing in this manner, the squaring rectangles assertion can be verified.

21. Suppose that $n = 2$, then $\Phi^2 = F_2\Phi + F_1 = \Phi + 1$, as verified in the previous exercise. Next,
$\Phi^3 = (\Phi^2)\Phi = (\Phi + 1)\Phi = \Phi^2 + \Phi = (\Phi + 1) + \Phi = 2\Phi + 1$, and since $F_3 = 2$ and $F_2 = 1$ so $\Phi^3 = F_3\Phi + F_2$. Next,
$\Phi^4 = (\Phi^3)\Phi = (2\Phi + 1)\Phi = 2\Phi^2 + \Phi = 2(\Phi + 1) + \Phi = 3\Phi + 2$
and since $F_4 = 3$ and $F_3 = 2$, one has $\Phi^4 = F_4\Phi + F_3$. Continuing it can be shown that $\Phi^n = F_n\Phi + F_{n-1}$.

SUPPLEMENTARY EXERCISES CHAPTER 2

1. The simple interest is $I = (25{,}000)(0.08)(30/12) = \$5000$.

3. The interest is $10{,}000e^{(0.07)(15)} - 10{,}000 = \$18{,}576.51$.

5. The present sum is $\dfrac{175{,}000}{e^{1.6}} = \$35{,}331.89$.

7. Here, $P = \$1500$, $n = 10$, and $i = 0.07$. The future value is

$$\text{FV} = \frac{(1 + 0.07)^{10} - 1}{0.07}(1500) = \frac{(1.07)^{10} - 1}{0.07}(1500) = \$20{,}724.67.$$

9. Here, $P = \$100$, $n = 120$, and $i = 0.005$. The present value is

$$\text{PV} = \frac{1 - (1 + 0.005)^{-120}}{0.005}(100) = \frac{1 - (1.005)^{-120}}{0.005}(100) = \$9007.35.$$

11. Using the amortization formula, the payment is

$$\frac{(0.06)(25{,}000)}{\left(1-\dfrac{1}{(1+0.06)^{10}}\right)}=\frac{1500}{1-(1.06)^{-10}}=\$3396.70.$$

Period	Beginning Balance	Interest	Amortized Payment (Less Interest)	Payment
1	$25,000.00	$1500.00	$1896.70	$3396.70
2	$23,103.30	$1386.20	$2010.50	$3396.70
3	$21,092.80	$1265.57	$2131.13	$3396.70
4	$18,961.67	$1137.70	$2259.00	$3396.70
5	$16,702.67	$1002.16	$2394.54	$3396.70
6	$14,308.13	$858.49	$2538.21	$3396.70
7	$11,769.92	$706.20	$2690.50	$3396.70
8	$9079.42	$544.77	$2851.94	$3396.70
9	$6227.48	$373.65	$3023.05	$3396.70
10	$3204.43	$192.27	$3204.43	$3396.70
Total		$8967.02*	$25,000.00	$3396.70

*Slight round-off error.

13. The common difference is 3, so the next three terms are 11, 14, and 17.

15. Let x_1 and x_2 be the first two numbers called. The third number is $x_1+x_2=x_3$. In terms of the original pair of numbers called, $x_3=2x_2+x_1$. The next numbers called in terms of the original pair are $3x_2+2x_1$, $5x_2+3x_1$, $8x_2+5x_1$, and $13x_2+8x_1$. Note that the coefficients on x_2, 1, 2, 3, 5, 8, 13, ... and on x_1, 1, 1, 2, 3, 5, 8, ... are Fibonacci numbers. Next suppose $\frac{x}{w}=\frac{y}{v}$, then $xv=wy$. Adding yv to each side results in $xv+yv=wy+yv$, which yields $v(x+y)=y(w+v)$ or $\frac{x+y}{w+v}=\frac{y}{v}$. The Golden Ratio results from ratios of successive pairs of Fibonacci numbers, so $\frac{13x_2}{8x_2}=\frac{13}{8}\approx\phi$. Also $\frac{8x_1}{5x_1}=\frac{8}{5}\approx\phi$. Therefore, $\frac{13x_2+8x_1}{8x_2+5x_1}=\frac{x_8}{x_7}\approx\phi$. The same logic applies to any

⟶

successive pairs of numbers called regardless of the values of x_1 and x_2, which explains how the Round Robin Party Game leads to an estimate for the Golden Ratio.

17. Answers vary but many should find examples where the Golden Mean is approximated.

CHAPTER 3

MATRIX ALGEBRA

EXERCISES 3.1

1. a) $\begin{bmatrix} 3 & -5 & 4 & | & 5 \\ 1 & 1 & -1 & | & 0 \\ 2 & 3 & -2 & | & 2 \end{bmatrix}$ b) $\begin{bmatrix} 1 & -3 & 5 & | & 19 \\ 2 & 4 & -1 & | & -5 \\ -1 & 5 & 3 & | & 3 \end{bmatrix}$

3. $5x_1 + 3x_2 = 13$
$4x_1 + 7x_2 = 18$

5. $x_1 + 2x_2 + 3x_3 = 4$
$2x_1 + x_2 + 2x_3 = 3$
$4x_1 + x_2 - 2x_3 = 9$

7. $x = 4, y = 3, z = 1$

9. Let $x = \#$ children tickets and $y = \#$ adult tickets. The equations are

$x + y = 2400$
$4x + 5y = 10{,}500$

The augmented matrix is $\begin{bmatrix} 1 & 1 & | & 2400 \\ 4 & 5 & | & 10{,}500 \end{bmatrix}$.

Solutions Manual to Accompany Finite Mathematics: Models and Applications, First Edition.
Carla C. Morris and Robert M. Stark.

11. The last row contains a row of zeroes that has a nonzero number on the RHS (right-hand side)

13. The new row 1 becomes [row 1 – 2 row 3] or

$$\begin{bmatrix}1 & 1 & 2 & 9\end{bmatrix} - \begin{bmatrix}0 & 0 & 2 & 2\end{bmatrix} = \begin{bmatrix}1 & 1 & 0 & 7\end{bmatrix}$$

The new row 2 becomes [row 2 – 3 row 3] or

$$\begin{bmatrix}0 & 1 & 3 & 7\end{bmatrix} - \begin{bmatrix}0 & 0 & 3 & 3\end{bmatrix} = \begin{bmatrix}0 & 1 & 0 & 4\end{bmatrix} \text{ as indicated.}$$

$$\left[\begin{array}{ccc|c}1 & 1 & 2 & 9\\0 & 1 & 3 & 7\\0 & 0 & 1 & 1\end{array}\right] \to \left[\begin{array}{ccc|c}1 & 1 & 0 & 7\\0 & 1 & 0 & 4\\0 & 0 & 1 & 1\end{array}\right]$$

15. First, perform the operations $R_2 \to R_2 - 2R_1$ and $R_3 \to R_3 + R_1$ as indicated.

$$\left[\begin{array}{ccc|c}1 & -3 & 5 & 19\\2 & 4 & -1 & -5\\-1 & 5 & 3 & 3\end{array}\right] \to \left[\begin{array}{ccc|c}1 & -3 & 5 & 19\\0 & 10 & -11 & -43\\0 & 2 & 8 & 22\end{array}\right]$$

Next, row 2 and row 3 are interchanged and row 2 becomes 1/2 the original (there are other options, but these fractions are simpler) to yield

$$\left[\begin{array}{ccc|c}1 & -3 & 5 & 19\\0 & 1 & 4 & 11\\0 & 10 & -11 & -43\end{array}\right].$$

Now $R_3 \to R_3 - 10R_2$ to yield $\left[\begin{array}{ccc|c}1 & -3 & 5 & 19\\0 & 1 & 4 & 11\\0 & 0 & -51 & -153\end{array}\right]$.

Next, $R_3 \to \ (-1/51)R_3$ yields $\left[\begin{array}{ccc|c}1 & -3 & 5 & 19\\0 & 1 & 4 & 11\\0 & 0 & 1 & 3\end{array}\right]$.

This indicates that $x_3 = 3$. The second row yields $x_2 + 4(3) = 11$ or $x_2 = -1$. Next, using these values in the first row $x_1 - 3(-1) + 5(3) = 19$ indicates that $x_1 = 1$. The ordered triple solution is $(1, -1, 3)$.

EXERCISES 3.2

1. The subscripts indicate the row and column for the entry.

⟶

a) $a_{12} = 2$ (first row, second column)

b) $a_{23} = -1$

c) $a_{13} = -3$

d) $a_{22} = 4$

3. The matrices dimensions are $\mathbf{A} = 2 \times 2$, $\mathbf{B} = 2 \times 2$, and $\mathbf{C} = 2 \times 3$.

a) To add matrices, their dimensions are equal, so $\mathbf{A} + \mathbf{B}$ is possible.

b) Since **B** and **C** are not of same dimension, $\mathbf{B} + 2\mathbf{C}$ is not possible.

c) Multiplication by a scalar does not alter dimension, so $5\mathbf{C}$ is possible.

d) $\mathbf{A} + \mathbf{B} - \mathbf{C}$ is not possible as the dimension of **C** differs from those of **A** and **B.**

e) $3\mathbf{B}$ has equal dimension with **B** while $6\mathbf{A}$ has equal dimension with **A**. Therefore, $3\mathbf{B} - 6\mathbf{A}$ is possible.

5. The matrices dimensions are $\mathbf{A} = 3 \times 2$, $\mathbf{B} = 3 \times 2$, and $\mathbf{C} = 3 \times 3$.

a) To add matrices, their dimensions are equal, so $\mathbf{A} + \mathbf{B}$ is possible.

b) Multiplication by a scalar does not alter dimension, so $4\mathbf{C}$ is possible.

c) Since **A** and **B** are of equal dimension, $2\mathbf{A} + \mathbf{B}$ is possible.

d) $3\mathbf{A} - 2\mathbf{C}$ is not possible as the dimension of **A** differs from that of **C.**

e) To add matrices, their dimensions are equal, so $\mathbf{B} + \mathbf{C}$ is not possible.

7. The valid matrix operations in Exercise 3.2.3 were in parts a), c), and e)

$$\mathbf{A}+\mathbf{B} = \begin{bmatrix} 10 & 25 \\ 74 & 30 \end{bmatrix} + \begin{bmatrix} 20 & 52 \\ 41 & 36 \end{bmatrix} = \begin{bmatrix} 30 & 77 \\ 115 & 66 \end{bmatrix}$$

$$5\mathbf{C} = 5\begin{bmatrix} 21 & 62 & -37 \\ 19 & 40 & -91 \end{bmatrix} = \begin{bmatrix} 105 & 310 & -185 \\ 95 & 200 & -455 \end{bmatrix}$$

$$3\mathbf{B} - 6\mathbf{A} = 3\begin{bmatrix} 20 & 52 \\ 41 & 36 \end{bmatrix} - 6\begin{bmatrix} 10 & 25 \\ 74 & 30 \end{bmatrix}$$

$$= \begin{bmatrix} 60 & 156 \\ 123 & 108 \end{bmatrix} - \begin{bmatrix} 60 & 150 \\ 444 & 180 \end{bmatrix} = \begin{bmatrix} 0 & 6 \\ -321 & -72 \end{bmatrix}$$

9. The valid operations in Exercise 3.2.5 are a), b), and c).

a) $\mathbf{A}+\mathbf{B}=\begin{bmatrix}4 & 1\\ 3 & 5\\ 6 & 1\end{bmatrix}+\begin{bmatrix}8 & -1\\ 12 & 3\\ 10 & 7\end{bmatrix}=\begin{bmatrix}12 & 0\\ 15 & 8\\ 16 & 8\end{bmatrix}$

b) $4\mathbf{C}=\mathbf{4}\begin{bmatrix}3 & 8 & 9\\ 2 & 0 & 0\\ 4 & 0 & 3\end{bmatrix}=\begin{bmatrix}12 & 32 & 36\\ 8 & 0 & 0\\ 16 & 0 & 12\end{bmatrix}$

c) $2\mathbf{A}+\mathbf{B}=2\begin{bmatrix}4 & 1\\ 3 & 5\\ 6 & 1\end{bmatrix}+\begin{bmatrix}8 & -1\\ 12 & 3\\ 10 & 7\end{bmatrix}=\begin{bmatrix}16 & 1\\ 18 & 13\\ 22 & 9\end{bmatrix}$

11. All the operations are valid in this case.

a) $\mathbf{A}+\mathbf{B}=\begin{bmatrix}1 & 0 & 2\\ 3 & 4 & 7\end{bmatrix}+\begin{bmatrix}-2 & 3 & 6\\ -5 & 1 & 4\end{bmatrix}=\begin{bmatrix}-1 & 3 & 8\\ -2 & 5 & 11\end{bmatrix}$

b) $2\mathbf{B}=2\begin{bmatrix}-2 & 3 & 6\\ -5 & 1 & 4\end{bmatrix}=\begin{bmatrix}-4 & 6 & 12\\ -10 & 2 & 8\end{bmatrix}$

c) $\mathbf{A}+3\mathbf{C}=\begin{bmatrix}1 & 0 & 2\\ 3 & 4 & 7\end{bmatrix}+3\begin{bmatrix}4 & 6 & -3\\ 2 & 8 & -9\end{bmatrix}=\begin{bmatrix}13 & 18 & -7\\ 9 & 28 & -20\end{bmatrix}$

d) $3\mathbf{A}-2\mathbf{B}+\mathbf{C}=3\begin{bmatrix}1 & 0 & 2\\ 3 & 4 & 7\end{bmatrix}-2\begin{bmatrix}-2 & 3 & 6\\ -5 & 1 & 4\end{bmatrix}+\begin{bmatrix}4 & 6 & -3\\ 2 & 8 & -9\end{bmatrix}$

$$=\begin{bmatrix}3 & 0 & 6\\ 9 & 12 & 21\end{bmatrix}-\begin{bmatrix}-4 & 6 & 12\\ -10 & 2 & 8\end{bmatrix}+\begin{bmatrix}4 & 6 & -3\\ 2 & 8 & -9\end{bmatrix}$$

$$=\begin{bmatrix}11 & 0 & -9\\ 21 & 18 & 4\end{bmatrix}$$

13. Since **A** is 1×4 and **B** is 4×1, the product is a 1×1 matrix.

$$\mathbf{AB}=[1\cdot(-2)+0\cdot(-1)+3\cdot 8+4\cdot 0]=[22]$$

Since **B** is 4×1 and **A** is 1×4, the product **BA** is a 4×4 matrix

$$\mathbf{BA}=\begin{bmatrix}-2 & 0 & -6 & -8\\ -1 & 0 & -3 & -4\\ 8 & 0 & 24 & 32\\ 0 & 0 & 0 & 0\end{bmatrix}$$

15. The dimensions for the matrices are $\mathbf{A}=2\times 2$, $\mathbf{B}=3\times 2$, and $\mathbf{C}=2\times 3$.

a) **AB** is not possible since the number of columns in **A** is not the same as the number of rows in **B** ⟶

b) **AC** is possible $(2 \times 2)(2 \times 3)$ and is 2×3

c) **BA** is possible $(3 \times 2)(2 \times 2)$ and is 3×2

d) **BC** is possible $(3 \times 2)(2 \times 3)$ and is 3×3

e) **CA** is not possible

f) **CB** is possible $(2 \times 3)(3 \times 2)$ and is 2×2

17. **AC**, **BA**, **BC**, and **CB** are possible

$$\mathbf{AC} = \begin{bmatrix} 0 & 2 \\ 4 & 6 \end{bmatrix} \begin{bmatrix} 1 & 2 & -3 \\ 9 & 4 & -1 \end{bmatrix} = \begin{bmatrix} 18 & 8 & -2 \\ 58 & 32 & -18 \end{bmatrix}$$

$$\mathbf{BA} = \begin{bmatrix} 2 & 5 \\ 4 & 3 \\ 1 & -2 \end{bmatrix} \begin{bmatrix} 0 & 2 \\ 4 & 6 \end{bmatrix} = \begin{bmatrix} 20 & 34 \\ 12 & 26 \\ -8 & -10 \end{bmatrix}$$

$$\mathbf{BC} = \begin{bmatrix} 2 & 5 \\ 4 & 3 \\ 1 & -2 \end{bmatrix} \begin{bmatrix} 1 & 2 & -3 \\ 9 & 4 & -1 \end{bmatrix} = \begin{bmatrix} 47 & 24 & -11 \\ 31 & 20 & -15 \\ -17 & -6 & -1 \end{bmatrix}$$

$$\mathbf{CB} = \begin{bmatrix} 1 & 2 & -3 \\ 9 & 4 & -1 \end{bmatrix} \begin{bmatrix} 2 & 5 \\ 4 & 3 \\ 1 & -2 \end{bmatrix} = \begin{bmatrix} 7 & 17 \\ 33 & 59 \end{bmatrix}$$

19. **A** is a (3×2) matrix

a) The zero matrix must have the same dimension as **A** for $\mathbf{A} + \mathbf{0}$ to be valid, so **0** is a (3×2) matrix.

b) The only requirement for the multiplication **A0** to be valid is that **0** has the same number of rows as the columns of **A**, so **0** is any $(2 \times n)$ matrix.

c) For **0A** to be valid, the number of columns of the zero matrix must be the same as the number of rows of **A**, so **0** is any $(m \times 3)$ matrix.

21. a) $$\mathbf{AA} = \begin{bmatrix} 3 & 0 & 1 \\ 0 & 1 & 2 \\ 1 & 2 & 4 \end{bmatrix} \begin{bmatrix} 3 & 0 & 1 \\ 0 & 1 & 2 \\ 1 & 2 & 4 \end{bmatrix} = \begin{bmatrix} 10 & 2 & 7 \\ 2 & 5 & 10 \\ 7 & 10 & 21 \end{bmatrix}.$$

b) It must be a square matrix (the same number of rows as columns).

23. a) The transpose of a column vector $(m \times 1)$ is a row vector $(1 \times m)$.

b) The transpose of a row vector $(1 \times m)$ is a column vector $(m \times 1)$

25. a) 5 pounds of cheese, 3 pounds of butter, and 2 pounds of dry milk on May 24, 2014 cost

$$\begin{bmatrix} 5 & 3 & 2 \end{bmatrix} \begin{bmatrix} 2.1128 \\ 2.1189 \\ 1.8143 \end{bmatrix} = \$20.5493$$

⟶

b) 4 pounds of cheese, 1 pound of butter, and 3 pounds of dry milk for each of the three weeks can be found using

$$\begin{bmatrix}4 & 1 & 3\end{bmatrix}\begin{bmatrix}2.0516 & 2.0234 & 2.0172\\ 2.1261 & 2.1857 & 2.1836\\ 1.8901 & 1.8427 & 1.8624\end{bmatrix}$$

$$= \begin{bmatrix}16.0028 & 15.8074 & 15.8396\end{bmatrix}$$

27. GPA is determined by the quality points being divided by the total of course credits. Find the quality points by multiplying the credits by the grades on the four point scale. There are 14 credits to a semester.

$$\begin{bmatrix}4 & 1 & 3 & 3 & 3\end{bmatrix}\begin{bmatrix}4.00\\ 3.67\\ 2.67\\ 3.00\\ 2.33\end{bmatrix} = [43.67]$$

$43.67/14 = 3.119$ semester GPA.

29. a) $$\begin{bmatrix}3 & -7 & 1 & 3\\ 2 & 3 & 1 & -2\\ 3 & 1 & 3 & 7\\ 1 & 5 & 2 & -8\end{bmatrix}\begin{bmatrix}x_1\\ x_2\\ x_3\\ x_4\end{bmatrix} = \begin{bmatrix}7\\ 0\\ 25\\ -19\end{bmatrix}$$

b) $$\begin{bmatrix}1 & 4 & 3 & 1\\ 2 & 3 & -1 & 5\\ 1 & 1 & -2 & 4\\ 3 & 2 & -1 & 3\end{bmatrix}\begin{bmatrix}w\\ x\\ y\\ z\end{bmatrix} = \begin{bmatrix}8\\ 7\\ 3\\ 10\end{bmatrix}$$

31. a) $$\begin{aligned} w + 5x + 2y + 2z &= 19\\ w + x + 3y \quad\quad &= 14\\ 2w + 3x - 2y + 5z &= 15\\ 3w + x \quad\quad + 2z &= 17\end{aligned}$$

b) $$\begin{aligned} 3x_1 + x_2 + 4x_3 + x_4 &= 12\\ x_1 + x_2 + 3x_3 + 2x_4 &= 13\\ x_1 + 2x_2 \quad\quad + 5x_4 &= 17\\ 2x_1 + 3x_2 + x_3 + x_4 &= 8\end{aligned}$$

33. a) $$\left[\begin{array}{cccc|c}1 & 5 & 2 & 2 & 19\\ 1 & 1 & 3 & 0 & 14\\ 2 & 3 & -2 & 5 & 15\\ 3 & 1 & 0 & 2 & 17\end{array}\right]$$

b) $$\left[\begin{array}{cccc|c}3 & 1 & 4 & 1 & 12\\ 1 & 1 & 3 & 2 & 13\\ 1 & 2 & 0 & 5 & 17\\ 2 & 3 & 1 & 1 & 8\end{array}\right]$$

EXERCISES 3.3

1. a) $\begin{bmatrix} 2 & 3 & | & 0 \\ 1 & 4 & | & 0 \end{bmatrix} \to \begin{bmatrix} 1 & 4 & | & 0 \\ 2 & 3 & | & 0 \end{bmatrix} \to \begin{bmatrix} 1 & 4 & | & 0 \\ 0 & -5 & | & 0 \end{bmatrix} \to \begin{bmatrix} 1 & 4 & | & 0 \\ 0 & 1 & | & 0 \end{bmatrix} \to \begin{bmatrix} 1 & 0 & | & 0 \\ 0 & 1 & | & 0 \end{bmatrix}$

Indicating the trivial solution $x_1 = 0$ and $x_2 = 0$.

b) $\begin{bmatrix} 1 & 3 & | & 0 \\ 2 & 6 & | & 0 \end{bmatrix} \to \begin{bmatrix} 1 & 3 & | & 0 \\ 0 & 0 & | & 0 \end{bmatrix}$

Indicating $x_1 = -3x_2$ and any value of x_2 determines x_1.

3. First, recall that elementary matrices are the result of performing a single row operation on an identity matrix.

a) Elementary $R_2 \to 9R_2$

b) Not elementary. Here, $R_3 \to 2R_3$ and $R_2 \to R_2 + 5R_1$

c) Elementary. Here, $R_2 \to 3R_2 + R_3$

d) Elementary. Here, $R_3 \to R_3 - 8R_2$

e) Elementary. Here, $R_2 \leftrightarrow R_1$

5. $\mathbf{A} = \begin{bmatrix} 10 & 3 & 7 & 5 \\ 12 & 8 & 1 & 0 \\ 7 & -4 & 2 & 6 \end{bmatrix}$

a) Eliminating the third row yields the submatrix $\begin{bmatrix} 10 & 3 & 7 & 5 \\ 12 & 8 & 1 & 0 \end{bmatrix}$.

b) Eliminating the second and fourth columns yields $\begin{bmatrix} 10 & 7 \\ 12 & 1 \\ 7 & 2 \end{bmatrix}$.

c) Eliminating the second row and second column yields $\begin{bmatrix} 10 & 7 & 5 \\ 7 & 2 & 6 \end{bmatrix}$.

d) Eliminating the first and third columns yields $\begin{bmatrix} 3 & 5 \\ 8 & 0 \\ -4 & 6 \end{bmatrix}$.

7. a) Row echelon form (but not reduced row echelon because of 5 in first row).

b) Row echelon form (but not reduced row echelon).

c) Not in row echelon form because leading 1 farther to right than in the third row.

d) Not in row echelon, the row of zeroes in second row is in the bottom row.

9. $\left[\begin{array}{cc|c} 2 & 1 & 7 \\ 4 & 1 & 9 \end{array}\right] \rightarrow \left[\begin{array}{cc|c} 1 & 1/2 & 7/2 \\ 4 & 1 & -19 \end{array}\right] \rightarrow \left[\begin{array}{cc|c} 1 & 1/2 & 7/2 \\ 0 & -1 & -5 \end{array}\right] \rightarrow \left[\begin{array}{cc|c} 1 & 1/2 & 7/2 \\ 0 & 1 & 5 \end{array}\right] \rightarrow$

$\left[\begin{array}{cc|c} 1 & 0 & 1 \\ 0 & 1 & 5 \end{array}\right]$ indicating $x_1 = 1$ and $x_2 = 5$.

11. $\left[\begin{array}{cc|c} 1 & 2 & 8 \\ 2 & 4 & 14 \end{array}\right] \rightarrow \left[\begin{array}{cc|c} 1 & 2 & 8 \\ 0 & 0 & -2 \end{array}\right]$ indicating no solution.

13. $\left[\begin{array}{ccc|c} 1 & -1 & -2 & -10 \\ 3 & 2 & -1 & 0 \\ 5 & 3 & 1 & 16 \end{array}\right] \rightarrow \left[\begin{array}{ccc|c} 1 & -1 & -2 & -10 \\ 0 & 5 & 5 & 30 \\ 0 & 8 & 11 & 66 \end{array}\right] \rightarrow \left[\begin{array}{ccc|c} 1 & -1 & -2 & -10 \\ 0 & 1 & 1 & 6 \\ 0 & 8 & 11 & 66 \end{array}\right] \rightarrow$

$\left[\begin{array}{ccc|c} 1 & -1 & -2 & -10 \\ 0 & 1 & 1 & 6 \\ 0 & 0 & 3 & 18 \end{array}\right] \rightarrow \left[\begin{array}{ccc|c} 1 & -1 & -2 & -10 \\ 0 & 1 & 1 & 6 \\ 0 & 0 & 1 & 6 \end{array}\right] \rightarrow \left[\begin{array}{ccc|c} 1 & -1 & 0 & 2 \\ 0 & 1 & 0 & 0 \\ 0 & 0 & 1 & 6 \end{array}\right] \rightarrow$

$\left[\begin{array}{ccc|c} 1 & 0 & 0 & 2 \\ 0 & 1 & 0 & 0 \\ 0 & 0 & 1 & 6 \end{array}\right]$ indicating the solution is $x_1 = 2$, $x_2 = 0$, and $x_3 = 6$.

15. $\left[\begin{array}{ccc|c} 1 & 2 & -3 & 5 \\ 3 & -1 & 1 & 0 \\ 2 & -1 & 9 & 3 \end{array}\right] \rightarrow \left[\begin{array}{ccc|c} 1 & 2 & -3 & 5 \\ 0 & -7 & 10 & -15 \\ 0 & -5 & 9 & -7 \end{array}\right] \rightarrow$

$\left[\begin{array}{ccc|c} 1 & 2 & -3 & 5 \\ 0 & 1 & -10/7 & 15/7 \\ 0 & -5 & 9 & -7 \end{array}\right] \rightarrow \left[\begin{array}{ccc|c} 1 & 2 & -3 & 5 \\ 0 & 1 & -10/7 & 15/7 \\ 0 & 0 & 13/7 & 26/7 \end{array}\right] \rightarrow$

$\left[\begin{array}{ccc|c} 1 & 2 & -3 & 5 \\ 0 & 1 & -10/7 & 15/7 \\ 0 & 0 & 1 & 2 \end{array}\right] \rightarrow \left[\begin{array}{ccc|c} 1 & 2 & 0 & 11 \\ 0 & 1 & 0 & 5 \\ 0 & 0 & 1 & 2 \end{array}\right] \rightarrow \left[\begin{array}{ccc|c} 1 & 0 & 0 & 1 \\ 0 & 1 & 0 & 5 \\ 0 & 0 & 1 & 2 \end{array}\right]$

indicating the solution is $x_1 = 1$, $x_2 = 5$, and $x_3 = 2$.

17. $\left[\begin{array}{cccc|c} 1 & 1 & 1 & 1 & 1 \\ 1 & -1 & -2 & 0 & 0 \\ 3 & 1 & 3 & -1 & 3 \\ 2 & -1 & 2 & 2 & 8 \end{array}\right] \rightarrow \left[\begin{array}{cccc|c} 1 & 1 & 1 & 1 & 1 \\ 0 & -2 & -3 & -1 & -1 \\ 0 & -2 & 0 & -4 & 0 \\ 0 & -3 & 0 & 0 & 6 \end{array}\right] \rightarrow$

$\left[\begin{array}{cccc|c} 1 & 1 & 1 & 1 & 1 \\ 0 & -3 & 0 & 0 & 6 \\ 0 & -2 & 0 & -4 & 0 \\ 0 & -2 & -3 & -1 & -1 \end{array}\right] \rightarrow \left[\begin{array}{cccc|c} 1 & 1 & 1 & 1 & 1 \\ 0 & 1 & 0 & 0 & -2 \\ 0 & -2 & 0 & -4 & 0 \\ 0 & -2 & -3 & -1 & -1 \end{array}\right] \rightarrow$

$\longrightarrow$

$$\left[\begin{array}{cccc|c} 1 & 1 & 1 & 1 & 1 \\ 0 & 1 & 0 & 0 & -2 \\ 0 & 0 & 0 & -4 & -4 \\ 0 & 0 & -3 & -1 & -5 \end{array}\right] \rightarrow \left[\begin{array}{cccc|c} 1 & 1 & 1 & 1 & 1 \\ 0 & 1 & 0 & 0 & -2 \\ 0 & 0 & -3 & -1 & -5 \\ 0 & 0 & 0 & -4 & -4 \end{array}\right] \rightarrow$$

$$\left[\begin{array}{cccc|c} 1 & 1 & 1 & 1 & 1 \\ 0 & 1 & 0 & 0 & -2 \\ 0 & 0 & -3 & -1 & -5 \\ 0 & 0 & 0 & 1 & 1 \end{array}\right] \rightarrow \left[\begin{array}{cccc|c} 1 & 1 & 1 & 1 & 1 \\ 0 & 1 & 0 & 0 & -2 \\ 0 & 0 & 1 & 1/3 & 5/3 \\ 0 & 0 & 0 & 1 & 1 \end{array}\right] \rightarrow$$

$$\left[\begin{array}{cccc|c} 1 & 1 & 1 & 0 & 0 \\ 0 & 1 & 0 & 0 & -2 \\ 0 & 0 & 1 & 0 & 4/3 \\ 0 & 0 & 0 & 1 & 1 \end{array}\right] \rightarrow \left[\begin{array}{cccc|c} 1 & 1 & 0 & 0 & -4/3 \\ 0 & 1 & 0 & 0 & -2 \\ 0 & 0 & 1 & 0 & 4/3 \\ 0 & 0 & 0 & 1 & 1 \end{array}\right] \rightarrow$$

$$\left[\begin{array}{cccc|c} 1 & 0 & 0 & 0 & 2/3 \\ 0 & 1 & 0 & 0 & -2 \\ 0 & 0 & 1 & 0 & 4/3 \\ 0 & 0 & 0 & 1 & 1 \end{array}\right] \rightarrow$$

indicating the solution is $x_1 = 2/3$, $x_2 = -2$, $x_3 = 4/3$, and $x_4 = 1$.

19. Let x_1 and x_2 be the advanced sales of children and adult tickets and x_3 the number of tickets sold at the door.

$$\begin{aligned} x_1 + x_2 + x_3 &= 1000 \\ 15x_1 + 20x_2 + 25x_3 &= 18{,}500 \\ 1/4x_1 \qquad\qquad - x_3 &= 0 \end{aligned}$$

$$\left[\begin{array}{ccc|c} 1 & 1 & 1 & 1000 \\ 15 & 20 & 25 & 18{,}500 \\ 1/4 & 0 & -1 & 0 \end{array}\right] \rightarrow \left[\begin{array}{ccc|c} 1 & 1 & 1 & 1000 \\ 0 & 5 & 10 & 3500 \\ 0 & -1/4 & -5/4 & -250 \end{array}\right] \rightarrow$$

$$\left[\begin{array}{ccc|c} 1 & 1 & 1 & 1000 \\ 0 & 1 & 2 & 700 \\ 0 & -1/4 & -5/4 & -250 \end{array}\right] \rightarrow \left[\begin{array}{ccc|c} 1 & 1 & 1 & 1000 \\ 0 & 1 & 2 & 700 \\ 0 & 0 & -3/4 & -75 \end{array}\right] \rightarrow$$

$$\left[\begin{array}{ccc|c} 1 & 1 & 1 & 1000 \\ 0 & 1 & 2 & 700 \\ 0 & 0 & 1 & 100 \end{array}\right] \rightarrow \left[\begin{array}{ccc|c} 1 & 1 & 0 & 900 \\ 0 & 1 & 0 & 500 \\ 0 & 0 & 1 & 100 \end{array}\right] \rightarrow \left[\begin{array}{ccc|c} 1 & 0 & 0 & 400 \\ 0 & 1 & 0 & 500 \\ 0 & 0 & 1 & 100 \end{array}\right] \rightarrow$$

$\longrightarrow$

There were 400 children's tickets and 500 adult tickets sold in advance. 100 tickets sold at the door.

21. Let $x = \#$ children tickets and $y = \#$ adult tickets. The equations are

$$x + y = 2400$$
$$4x + 5y = 10{,}500$$

The augmented matrix is $\left[\begin{array}{cc|c} 1 & 1 & 2400 \\ 4 & 5 & 10{,}500 \end{array}\right]$.

Next, by using Gauss Jordan, one gets the following matrix:

$$\left[\begin{array}{cc|c} 1 & 1 & 2400 \\ 4 & 5 & 10{,}500 \end{array}\right] \rightarrow \left[\begin{array}{cc|c} 1 & 1 & 2400 \\ 0 & 1 & 900 \end{array}\right] \rightarrow \left[\begin{array}{cc|c} 1 & 0 & 1500 \\ 0 & 1 & 900 \end{array}\right]$$

Therefore, 1500 children tickets and 900 adult tickets were sold.

23. Let x_1 be the amount of Food A, x_2 the amount of Food B, and x_3 the amount of Food C.

The system is

$$50x_1 + 100x_2 + 125x_3 = 525$$
$$20x_1 + 25x_2 + 18x_3 = 133$$
$$2x_2 + 3x_3 = 9$$

$$\left[\begin{array}{ccc|c} 50 & 100 & 125 & 525 \\ 20 & 25 & 18 & 133 \\ 0 & 2 & 3 & 9 \end{array}\right] \rightarrow \left[\begin{array}{ccc|c} 1 & 2 & 5/2 & 21/2 \\ 0 & 2 & 3 & 9 \\ 20 & 25 & 18 & 133 \end{array}\right] \rightarrow$$

$$\left[\begin{array}{ccc|c} 1 & 2 & 5/2 & 21/2 \\ 0 & 2 & 3 & 9 \\ 0 & -15 & -32 & -77 \end{array}\right] \rightarrow \left[\begin{array}{ccc|c} 1 & 2 & 5/2 & 21/2 \\ 0 & 1 & 3/2 & 9/2 \\ 0 & -15 & -32 & -77 \end{array}\right] \rightarrow$$

$$\left[\begin{array}{ccc|c} 1 & 2 & 5/2 & 21/2 \\ 0 & 1 & 3/2 & 9/2 \\ 0 & 0 & -19/2 & -19/2 \end{array}\right] \rightarrow \left[\begin{array}{ccc|c} 1 & 2 & 5/2 & 21/2 \\ 0 & 1 & 3/2 & 9/2 \\ 0 & 0 & 1 & 1 \end{array}\right] \rightarrow$$

$$\left[\begin{array}{ccc|c} 1 & 2 & 0 & 8 \\ 0 & 1 & 0 & 3 \\ 0 & 0 & 1 & 1 \end{array}\right] \rightarrow \left[\begin{array}{ccc|c} 1 & 0 & 0 & 2 \\ 0 & 1 & 0 & 3 \\ 0 & 0 & 1 & 1 \end{array}\right] \rightarrow$$

Therefore, two ounces of Food A, three ounces of Food B, and one ounce of Food C are combined.

EXERCISES 3.4

1. $\mathbf{AB} = \begin{bmatrix} 3 & 1 \\ 5 & 2 \end{bmatrix} \begin{bmatrix} 2 & -1 \\ -5 & 3 \end{bmatrix} = \begin{bmatrix} 1 & 0 \\ 0 & 1 \end{bmatrix} = \mathbf{I}$

$\mathbf{BA} = \begin{bmatrix} 2 & -1 \\ -5 & 3 \end{bmatrix} \begin{bmatrix} 3 & 1 \\ 5 & 2 \end{bmatrix} = \begin{bmatrix} 1 & 0 \\ 0 & 1 \end{bmatrix} = \mathbf{I}$

Since the matrix products result in an identity matrix, the matrices are inverses.

3. $\mathbf{AB} = \begin{bmatrix} 3 & 5 & 9 \\ 1 & 3 & 6 \\ 0 & 1 & 0 \end{bmatrix} \begin{bmatrix} 3 & 0 & -5 \\ -1 & 2 & 4 \\ 1 & 0 & 0 \end{bmatrix} = \begin{bmatrix} 13 & 10 & 5 \\ 6 & 6 & 7 \\ -1 & 2 & 4 \end{bmatrix} \neq \mathbf{I}$

$\mathbf{BA} = \begin{bmatrix} 3 & 0 & -5 \\ -1 & 2 & 4 \\ 1 & 0 & 0 \end{bmatrix} \begin{bmatrix} 3 & 5 & 9 \\ 1 & 3 & 6 \\ 0 & 1 & 0 \end{bmatrix} = \begin{bmatrix} 9 & 10 & 27 \\ -1 & 5 & 3 \\ 3 & 5 & 9 \end{bmatrix} \neq \mathbf{I}$

The matrices are not inverses as their products are not an identity matrix.

5. a) $\begin{bmatrix} 3 & 4 \\ 6 & 8 \end{bmatrix}$ The second row is a multiple of the first row.

b) $\begin{bmatrix} 2 & 3 & 5 \\ 0 & 2 & 3 \\ 4 & 6 & 10 \end{bmatrix}$ The third row is a multiple of the first row.

c) $\begin{bmatrix} 1 & 0 & 6 & 5 \\ 2 & 3 & 3 & 4 \\ 0 & 0 & 0 & 0 \\ 1 & 1 & 7 & 2 \end{bmatrix}$ The third row is all zeroes.

d) $\begin{bmatrix} 1 & 1 & 3 \\ 2 & 4 & 6 \\ 3 & 0 & 9 \end{bmatrix}$ The third column is a multiple of the first column.

7. a) $\mathbf{A}^{-1}\mathbf{A} = \begin{bmatrix} 5/4 & 1/4 & 3/4 \\ 1/4 & 5/4 & 3/4 \\ 1/2 & 1/2 & 3/2 \end{bmatrix} \begin{bmatrix} 1 & 0 & -1/2 \\ 0 & 1 & -1/2 \\ -1/3 & -1/3 & 1 \end{bmatrix}$

$= \begin{bmatrix} 1 & 0 & 0 \\ 0 & 1 & 0 \\ 0 & 0 & 1 \end{bmatrix} = \mathbf{I}$

⟶

$$\mathbf{AA}^{-1} = \begin{bmatrix} 1 & 0 & -1/2 \\ 0 & 1 & -1/2 \\ -1/3 & -1/3 & 1 \end{bmatrix} \begin{bmatrix} 5/4 & 1/4 & 3/4 \\ 1/4 & 5/4 & 3/4 \\ 1/2 & 1/2 & 3/2 \end{bmatrix}$$

$$= \begin{bmatrix} 1 & 0 & 0 \\ 0 & 1 & 0 \\ 0 & 0 & 1 \end{bmatrix} = \mathbf{I}$$

b) Using the computational technique,

$$\left[\begin{array}{ccc|ccc} 1 & 0 & -1/2 & 1 & 0 & 0 \\ 0 & 1 & -1/2 & 0 & 1 & 0 \\ -1/3 & -1/3 & 1 & 0 & 0 & 1 \end{array}\right] \to \left[\begin{array}{ccc|ccc} 1 & 0 & -1/2 & 1 & 0 & 0 \\ 0 & 1 & -1/2 & 0 & 1 & 0 \\ 0 & -1/3 & 5/6 & 1/3 & 0 & 1 \end{array}\right]$$

$$\left[\begin{array}{ccc|ccc} 1 & 0 & -1/2 & 1 & 0 & 0 \\ 0 & 1 & -1/2 & 0 & 1 & 0 \\ 0 & 0 & 2/3 & 1/3 & 1/3 & 1 \end{array}\right] \to \left[\begin{array}{ccc|ccc} 1 & 0 & -1/2 & 1 & 0 & 0 \\ 0 & 1 & -1/2 & 0 & 1 & 0 \\ 0 & 0 & 1 & 1/2 & 1/2 & 3/2 \end{array}\right]$$

$$\left[\begin{array}{ccc|ccc} 1 & 0 & 0 & 5/4 & 1/4 & 3/4 \\ 0 & 1 & 0 & 1/4 & 5/4 & 3/4 \\ 0 & 0 & 1 & 1/2 & 1/2 & 3/2 \end{array}\right]$$ indicating that the inverse is

$$\begin{bmatrix} 5/4 & 1/4 & 3/4 \\ 1/4 & 5/4 & 3/4 \\ 1/2 & 1/2 & 3/2 \end{bmatrix}$$ which is in agreement with the earlier result.

9. $$\left[\begin{array}{ccc|ccc} 3 & 3 & -1 & 1 & 0 & 0 \\ -2 & -2 & 1 & 0 & 1 & 0 \\ -4 & -5 & 2 & 0 & 0 & 1 \end{array}\right] \to \left[\begin{array}{ccc|ccc} 1 & 1 & -1/3 & 1/3 & 0 & 0 \\ -2 & -2 & 1 & 0 & 1 & 0 \\ -4 & -5 & 2 & 0 & 0 & 1 \end{array}\right] \to$$

$$\left[\begin{array}{ccc|ccc} 1 & 1 & -1/3 & 1/3 & 0 & 0 \\ 0 & 0 & 1/3 & 2/3 & 1 & 0 \\ 0 & -1 & 2/3 & 4/3 & 0 & 1 \end{array}\right] \to \left[\begin{array}{ccc|ccc} 1 & 1 & -1/3 & 1/3 & 0 & 0 \\ 0 & -1 & 2/3 & 4/3 & 0 & 1 \\ 0 & 0 & 1 & 2 & 3 & 0 \end{array}\right] \to$$

$$\left[\begin{array}{ccc|ccc} 1 & 1 & -1/3 & 1/3 & 0 & 0 \\ 0 & 1 & -2/3 & -4/3 & 0 & -1 \\ 0 & 0 & 1 & 2 & 3 & 0 \end{array}\right] \to \left[\begin{array}{ccc|ccc} 1 & 1 & 0 & 1 & 1 & 0 \\ 0 & 1 & 0 & 0 & 2 & -1 \\ 0 & 0 & 1 & 2 & 3 & 0 \end{array}\right]$$

$$\left[\begin{array}{ccc|ccc} 1 & 0 & 0 & 1 & -1 & 1 \\ 0 & 1 & 0 & 0 & 2 & -1 \\ 0 & 0 & 1 & 2 & 3 & 0 \end{array}\right]$$ indicating $\mathbf{A}^{-1} = \begin{bmatrix} 1 & -1 & 1 \\ 0 & 2 & -1 \\ 2 & 3 & 0 \end{bmatrix}$

11. $$\left[\begin{array}{ccc|ccc}1&0&2&1&0&0\\3&1&0&0&1&0\\0&1&4&0&0&1\end{array}\right]\to\left[\begin{array}{ccc|ccc}1&0&2&1&0&0\\0&1&-6&-3&1&0\\0&1&4&0&0&1\end{array}\right]\to$$

$$\left[\begin{array}{ccc|ccc}1&0&2&1&0&0\\0&1&-6&-3&1&0\\0&0&10&3&-1&1\end{array}\right]\to\left[\begin{array}{ccc|ccc}1&0&2&1&0&0\\0&1&-6&-3&1&0\\0&0&1&0.3&-0.1&0.1\end{array}\right]\to$$

$$\left[\begin{array}{ccc|ccc}1&0&0&0.4&0.2&-0.2\\0&1&0&-1.2&0.4&0.6\\0&0&1&0.3&-0.1&0.1\end{array}\right]$$

Therefore, $\mathbf{A}^{-1}=\begin{bmatrix}0.4&0.2&-0.2\\-1.2&0.4&0.6\\0.3&-0.1&0.1\end{bmatrix}$

13. Let x = number of nickels, y = number of dimes, and z = number of quarters.

a) The system of equations is

$$\begin{aligned}x+\ \ y+\ \ z&=14\\x-\ \ y+\ \ z&=2\\5x+10y+25z&=160\end{aligned}$$

b) The system in matrix form is:

$$\begin{bmatrix}1&1&1\\1&-1&1\\5&10&25\end{bmatrix}\begin{bmatrix}x\\y\\z\end{bmatrix}=\begin{bmatrix}14\\2\\160\end{bmatrix}$$

c) The inverse of the coefficients matrix is determined as:

$$\left[\begin{array}{ccc|ccc}1&1&1&1&0&0\\1&-1&1&0&1&0\\5&10&25&0&0&1\end{array}\right]\to\left[\begin{array}{ccc|ccc}1&1&1&1&0&0\\0&-2&0&-1&1&0\\0&5&20&-5&0&1\end{array}\right]\to$$

$$\left[\begin{array}{ccc|ccc}1&1&1&1&0&0\\0&1&0&1/2&-1/2&0\\0&5&20&-5&0&1\end{array}\right]\to$$

$$\left[\begin{array}{ccc|ccc}1&1&1&1&0&0\\0&1&0&1/2&-1/2&0\\0&0&20&-15/2&5/2&1\end{array}\right]\to$$

⟶

$$\left[\begin{array}{ccc|ccc} 1 & 1 & 1 & 1 & 0 & 0 \\ 0 & 1 & 0 & 1/2 & -1/2 & 0 \\ 0 & 0 & 1 & -3/8 & 1/8 & 1/20 \end{array}\right] \rightarrow$$

$$\left[\begin{array}{ccc|ccc} 1 & 1 & 0 & 11/8 & -1/8 & -1/20 \\ 0 & 1 & 0 & 1/2 & -1/2 & 0 \\ 0 & 0 & 1 & -3/8 & 1/8 & 1/20 \end{array}\right] \rightarrow$$

$$\left[\begin{array}{ccc|ccc} 1 & 0 & 0 & 7/8 & 3/8 & -1/20 \\ 0 & 1 & 0 & 1/2 & -1/2 & 0 \\ 0 & 0 & 1 & -3/8 & 1/8 & 1/20 \end{array}\right]$$

Now, the solution is

$$\begin{bmatrix} 7/8 & 3/8 & -1/20 \\ 1/2 & -1/2 & 0 \\ -3/8 & 1/8 & 1/20 \end{bmatrix} \begin{bmatrix} 14 \\ 2 \\ 160 \end{bmatrix} = \begin{bmatrix} 5 \\ 6 \\ 3 \end{bmatrix}$$

Therefore, there are five nickels, six dimes, and three quarters.

15. Letting x = amount in stocks, y = amount in bonds, z = amount in market funds. The system of equations in matrix form is:

$$\begin{bmatrix} 1 & 1 & 1 \\ 1 & 0 & -2 \\ 0 & 1 & -1 \end{bmatrix} \begin{bmatrix} x \\ y \\ z \end{bmatrix} = \begin{bmatrix} 18{,}000 \\ 0 \\ 2000 \end{bmatrix}$$

The inverse of the coefficients matrix is: $\begin{bmatrix} 1/2 & 1/2 & -1/2 \\ 1/4 & -1/4 & 3/4 \\ 1/4 & -1/4 & -1/4 \end{bmatrix}$

The solution is: $\begin{bmatrix} 1/2 & 1/2 & -1/2 \\ 1/4 & -1/4 & 3/4 \\ 1/4 & -1/4 & -1/4 \end{bmatrix} \begin{bmatrix} 18{,}000 \\ 0 \\ 2000 \end{bmatrix} = \begin{bmatrix} 8000 \\ 6000 \\ 4000 \end{bmatrix}$

Therefore, \$8000 was invested in stocks, \$6000 in bonds, and \$4000 in money market funds.

17. Because the encoding matrix is 3×3, a $3 \times n$ matrix is needed for the matrix multiplication.

$\longrightarrow$

F	O	U	R	_	S	C	O	R	E	_	A	N	D	_	S	E	V	E	N	_
6	15	21	18	27	19	3	15	18	5	27	1	14	4	27	19	5	22	5	14	27

Y	E	A	R	S	_	A	G	O
25	5	1	18	19	27	1	7	15

Complete the matrix column by column as:

$$\begin{bmatrix} 6 & 18 & 3 & 5 & 14 & 19 & 5 & 25 & 18 & 1 \\ 15 & 27 & 15 & 27 & 4 & 5 & 14 & 5 & 19 & 7 \\ 21 & 19 & 18 & 1 & 27 & 22 & 27 & 1 & 27 & 15 \end{bmatrix}$$

To encode the message, multiply matrices.

$$\begin{bmatrix} 3 & 4 & 1 \\ 1 & 1 & 0 \\ 2 & 3 & 2 \end{bmatrix}\begin{bmatrix} 6 & 18 & 3 & 5 & 14 & 19 & 5 & 25 & 18 & 1 \\ 15 & 27 & 15 & 27 & 4 & 5 & 14 & 5 & 19 & 7 \\ 21 & 19 & 18 & 1 & 27 & 22 & 27 & 1 & 27 & 15 \end{bmatrix}$$
$$= \begin{bmatrix} 99 & 181 & 87 & 124 & 85 & 99 & 98 & 96 & 157 & 46 \\ 21 & 45 & 18 & 32 & 18 & 24 & 19 & 30 & 37 & 8 \\ 99 & 155 & 87 & 93 & 94 & 97 & 106 & 67 & 147 & 53 \end{bmatrix}$$

The encoded message is:

99 21 99 181 45 155 87 18 87 124 32 93 85 18 94
99 24 97 98 19 106 96 30 67 157 37 147 46 8 53

19. Decoding the message requires the inverse of the encoding key. The inverse of the key is $\begin{bmatrix} -2 & 5 & 1 \\ 2 & -4 & -1 \\ -1 & 1 & 1 \end{bmatrix}$.

The matrix multiplication is:

$$\begin{bmatrix} -2 & 5 & 1 \\ 2 & -4 & -1 \\ -1 & 1 & 1 \end{bmatrix}\begin{bmatrix} 99 & 181 & 87 & 124 & 85 & 99 & 98 & 96 & 157 & 46 \\ 21 & 45 & 18 & 32 & 18 & 24 & 19 & 30 & 37 & 8 \\ 99 & 155 & 87 & 93 & 94 & 97 & 106 & 67 & 147 & 53 \end{bmatrix}$$
$$= \begin{bmatrix} 6 & 18 & 3 & 5 & 14 & 19 & 5 & 25 & 18 & 1 \\ 15 & 27 & 15 & 27 & 4 & 5 & 14 & 5 & 19 & 7 \\ 21 & 19 & 18 & 1 & 27 & 22 & 27 & 1 & 27 & 15 \end{bmatrix}$$

revealing the message: FOUR SCORE AND SEVEN YEARS AGO

21. Here, $\mathbf{D} = (\mathbf{I} - \mathbf{A})\mathbf{X}$. Therefore,

$$\left(\begin{bmatrix} 1 & 0 & 0 \\ 0 & 1 & 0 \\ 0 & 0 & 1 \end{bmatrix} - \begin{bmatrix} 0.2 & 0.5 & 0.1 \\ 0.0 & 0.3 & 0.2 \\ 0.5 & 0.0 & 0.4 \end{bmatrix}\right)\begin{bmatrix} 30 \\ 20 \\ 50 \end{bmatrix}$$

$$= \begin{bmatrix} 0.8 & -0.5 & -0.1 \\ 0.0 & 0.7 & -0.2 \\ -0.5 & 0.0 & 0.6 \end{bmatrix}\begin{bmatrix} 30 \\ 20 \\ 50 \end{bmatrix} = \begin{bmatrix} 9 \\ 4 \\ 15 \end{bmatrix}$$

23. It is useful to imagine the technology matrix as

$$\begin{array}{c} \quad\quad\quad\textbf{outputs} \\ \textbf{inputs}\begin{bmatrix} \quad & \quad & \quad \\ & & \\ & & \end{bmatrix} \end{array}$$

The technology matrix $\mathbf{A}$ here is

$$\mathbf{A} = \begin{bmatrix} 0.10 & 0.40 & 0.35 \\ 0.30 & 0.25 & 0.15 \\ 0.20 & 0.15 & 0.35 \end{bmatrix} \text{ and } \mathbf{I} - \mathbf{A} = \begin{bmatrix} 0.90 & -0.40 & -0.35 \\ -0.30 & 0.75 & -0.15 \\ -0.20 & -0.15 & 0.65 \end{bmatrix}$$

Next, one determines $(\mathbf{I} - \mathbf{A})^{-1} \approx \begin{bmatrix} 1.787 & 1.201 & 1.239 \\ 0.865 & 1.979 & 0.922 \\ 0.749 & 0.826 & 2.133 \end{bmatrix}$

Finally, the production schedule is determined by $\mathbf{X} = (\mathbf{I} - \mathbf{A})^{-1}\mathbf{D}$

$$\begin{bmatrix} 1.787 & 1.201 & 1.239 \\ 0.865 & 1.979 & 0.922 \\ 0.749 & 0.826 & 2.133 \end{bmatrix}\begin{bmatrix} 25{,}000 \\ 10{,}000 \\ 15{,}000 \end{bmatrix} = \begin{bmatrix} 75{,}270 \\ 55{,}245 \\ 58{,}980 \end{bmatrix}$$

SUPPLEMENTARY EXERCISES CHAPTER 3

1. a) $\mathbf{AB}$ is defined, if $n = p$.

b) $\mathbf{A} + \mathbf{B}$ is defined, if $m = p$ and $n = r$.

c) $\mathbf{AB}$ and $\mathbf{BA}$ are defined if $n = p$ and $r = m$.

3. $\mathbf{AB} = \begin{bmatrix} 1 & 3 & 2 & 1 \\ 9 & 2 & 0 & 1 \\ 5 & 6 & 8 & 3 \\ 4 & 2 & 7 & 6 \end{bmatrix}\begin{bmatrix} 0 & 1 & 3 & -3 \\ 5 & 4 & 9 & -1 \\ 3 & 0 & 7 & -2 \\ 2 & 1 & 0 & 4 \end{bmatrix} = \begin{bmatrix} 23 & 14 & 44 & -6 \\ 12 & 18 & 45 & -25 \\ 60 & 32 & 125 & -25 \\ 43 & 18 & 79 & -4 \end{bmatrix}$

5. $\mathbf{A}^2 = \begin{bmatrix} 2 & 3 & 1 \\ 1 & 4 & 0 \\ 6 & 0 & 5 \end{bmatrix}\begin{bmatrix} 2 & 3 & 1 \\ 1 & 4 & 0 \\ 6 & 0 & 5 \end{bmatrix} = \begin{bmatrix} 13 & 18 & 7 \\ 6 & 19 & 1 \\ 42 & 18 & 31 \end{bmatrix}$

7. $\left[\begin{array}{cc|c} 3 & 4 & 7 \\ 1 & -2 & -1 \\ 2 & 3 & 5 \end{array}\right] \rightarrow \left[\begin{array}{cc|c} 1 & -1 & -1 \\ 3 & 4 & 7 \\ 2 & 3 & 5 \end{array}\right] \rightarrow \left[\begin{array}{cc|c} 1 & -2 & -1 \\ 0 & 10 & 10 \\ 0 & 7 & 7 \end{array}\right] \rightarrow \left[\begin{array}{cc|c} 1 & -2 & -1 \\ 0 & 1 & 1 \\ 0 & 7 & 7 \end{array}\right] \rightarrow$

$\left[\begin{array}{cc|c} 1 & -2 & -1 \\ 0 & 1 & 1 \\ 0 & 0 & 0 \end{array}\right] \rightarrow \left[\begin{array}{cc|c} 1 & 0 & 1 \\ 0 & 1 & 1 \\ 0 & 0 & 0 \end{array}\right] \rightarrow$ indicating $x_1 = 1$ and $x_2 = 1$

9. $\left[\begin{array}{cccc|c} 2 & 3 & 0 & 1 & 18 \\ 0 & -2 & 3 & -2 & -5 \\ 0 & 1 & 1 & 0 & 18 \\ 2 & 0 & 3 & -1 & 10 \end{array}\right] \rightarrow \left[\begin{array}{cccc|c} 1 & 3/2 & 0 & 1/2 & 9 \\ 0 & -2 & 3 & -2 & -5 \\ 0 & 1 & 1 & 0 & 18 \\ 2 & 0 & 3 & -1 & 10 \end{array}\right] \rightarrow$

$\left[\begin{array}{cccc|c} 1 & 3/2 & 0 & 1/2 & 9 \\ 0 & 1 & 1 & 0 & 18 \\ 0 & -2 & 3 & -2 & -5 \\ 2 & 0 & 3 & -1 & 10 \end{array}\right] \rightarrow \left[\begin{array}{cccc|c} 1 & 3/2 & 0 & 1/2 & 9 \\ 0 & 1 & 1 & 0 & 18 \\ 0 & -2 & 3 & -2 & -5 \\ 0 & -3 & 3 & -2 & 10 \end{array}\right]$

$\left[\begin{array}{cccc|c} 1 & 3/2 & 0 & 1/2 & 9 \\ 0 & 1 & 1 & 0 & 18 \\ 0 & 0 & 5 & -2 & 31 \\ 0 & 0 & 6 & -2 & 46 \end{array}\right] \rightarrow \left[\begin{array}{cccc|c} 1 & 3/2 & 0 & 1/2 & 9 \\ 0 & 1 & 1 & 0 & 18 \\ 0 & 0 & 1 & -2/5 & 31/5 \\ 0 & 0 & 6 & -2 & 46 \end{array}\right]$

$\left[\begin{array}{cccc|c} 1 & 3/2 & 0 & 1/2 & 9 \\ 0 & 1 & 1 & 0 & 18 \\ 0 & 0 & 1 & -2/5 & 31/5 \\ 0 & 0 & 0 & 2/5 & 46/5 \end{array}\right] \rightarrow \left[\begin{array}{cccc|c} 1 & 3/2 & 0 & 1/2 & 9 \\ 0 & 1 & 1 & 0 & 18 \\ 0 & 0 & 1 & -2/5 & 31/5 \\ 0 & 0 & 0 & 1 & 22 \end{array}\right]$

$\left[\begin{array}{cccc|c} 1 & 3/2 & 0 & 0 & -2 \\ 0 & 1 & 1 & 0 & 18 \\ 0 & 0 & 1 & 0 & 15 \\ 0 & 0 & 0 & 1 & 22 \end{array}\right] \rightarrow \left[\begin{array}{cccc|c} 1 & 3/2 & 0 & 0 & -2 \\ 0 & 1 & 0 & 0 & 3 \\ 0 & 0 & 1 & 0 & 15 \\ 0 & 0 & 0 & 1 & 22 \end{array}\right] \rightarrow$

$\left[\begin{array}{cccc|c} 1 & 0 & 0 & 0 & -13/2 \\ 0 & 1 & 0 & 0 & 3 \\ 0 & 0 & 1 & 0 & 15 \\ 0 & 0 & 0 & 1 & 22 \end{array}\right]$

⟶

Therefore, $x_1 = -13/2$, $x_2 = 3$, $x_3 = 15$, and $x_4 = 22$.

11. $\begin{bmatrix} 4p & 0 \\ 3p & 2p \end{bmatrix} + \begin{bmatrix} 6q & 1q \\ 2q & -q \end{bmatrix} = \begin{bmatrix} 16 & 2 \\ 7 & 0 \end{bmatrix}$. The first row entries are enough to solve. If $4p + 6q = 16$ and $0 + 1q = 2$, the solution is $p = 1$ and $q = 2$.

13. The augmented matrix is $\left[\begin{array}{ccccc|c} 4 & -11 & 5 & 3 & 3 & 5 \\ 1 & -1 & 1 & -2 & 4 & 10 \\ 3 & 1 & 2 & -3 & 1 & 4 \\ 1 & 1 & 3 & 2 & 2 & 10 \\ 1 & 0 & 5 & 0 & 3 & 7 \end{array}\right]$.

15. The receipts can be calculated by using the product of the ticket prices and their numbers.

$$\begin{bmatrix} 25 & 15 & 12 \end{bmatrix} \begin{bmatrix} 2000 & 2500 \\ 3000 & 2500 \\ 5000 & 4000 \end{bmatrix} = \begin{bmatrix} 155{,}000 & 163{,}000 \end{bmatrix}$$

The receipts were \$155,000 for the first performance and \$163,000 for the second for a weekend total of \$318,000.

17. The system of equations is

$$\begin{aligned} x_1 + x_2 + x_3 &= 6 \\ x_2 - x_3 &= -1 \\ x_1 - x_2 + x_3 &= 0 \end{aligned}$$

The solution is $x_1 = -1$, $x_2 = 3$, and $x_3 = 4$.

19. The matrix $\mathbf{I} - \mathbf{A} = \begin{bmatrix} 0.9 & -0.5 & -0.2 \\ -0.3 & 0.7 & -0.3 \\ -0.6 & 0.0 & 0.9 \end{bmatrix}$. Next, find the inverse.

$$(\mathbf{I} - \mathbf{A})^{-1} = \begin{bmatrix} \frac{315}{129} & \frac{225}{129} & \frac{145}{129} \\ \frac{75}{43} & \frac{115}{43} & \frac{55}{43} \\ \frac{70}{43} & \frac{50}{43} & \frac{80}{43} \end{bmatrix}$$

Finally, the production schedule is

$$\begin{bmatrix} \frac{315}{129} & \frac{225}{129} & \frac{145}{129} \\ \frac{75}{43} & \frac{115}{43} & \frac{55}{43} \\ \frac{70}{43} & \frac{50}{43} & \frac{80}{43} \end{bmatrix} \begin{bmatrix} 5 \\ 14 \\ 3 \end{bmatrix} = \begin{bmatrix} 40 \\ 50 \\ 30 \end{bmatrix}.$$

CHAPTER 4

LINEAR PROGRAMMING – GRAPHICAL SOLUTIONS

EXERCISES 4.1

1. a) $5x + 2y > 10$ is a strict inequality in two variables.

b) $x + 25y \leq 50$ is not a strict inequality.

c) $x_1 + 3x_2 < 9$ is a strict inequality in two variables.

3. To graph $x + 5 < 4$, rewrite the inequality as $x < -1$. As a strict inequality $x = -1$ is a dashed line. Then the proper half-plane is shaded.

⟶

Solutions Manual to Accompany Finite Mathematics: Models and Applications, First Edition.
Carla C. Morris and Robert M. Stark.

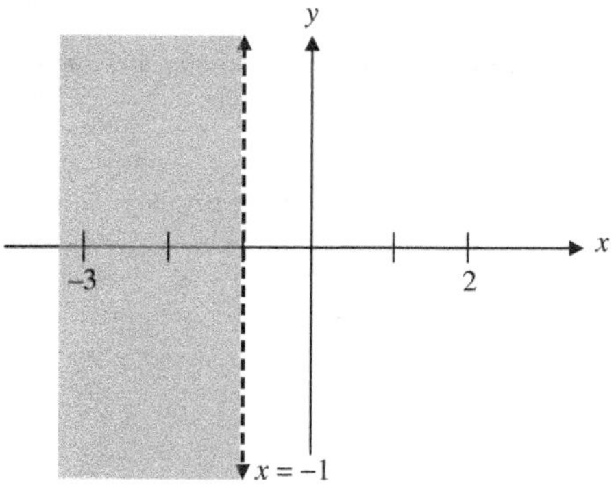

5. For $2x + 3y > 12$, the line $2x + 3y = 12$ is drawn as a dashed line and the appropriate half-plane shaded.

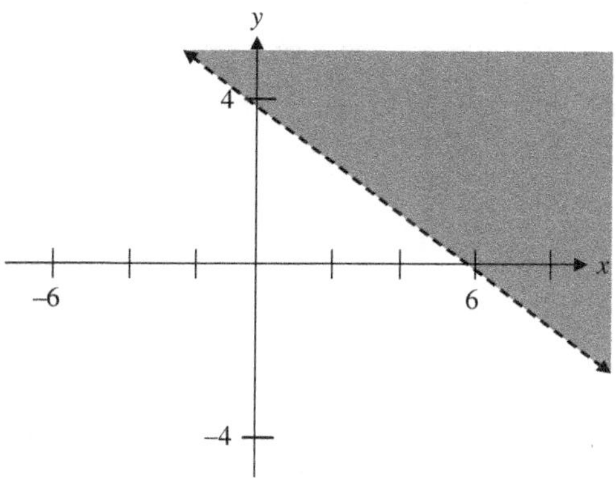

7. For the strict inequality $x < 3$, the line $x = 3$ is drawn as a dashed line and the proper half-plane is shaded.

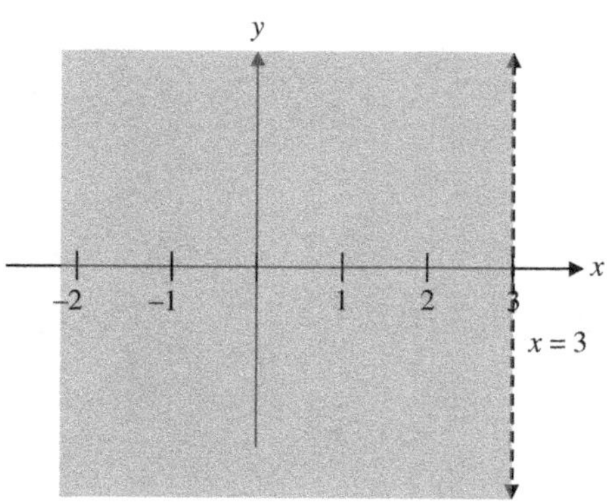

9. For the inequality $3x \leq 2y$, the line $2y = 3x$ or $y = (3/2)x$ is drawn as a solid line and the proper half-plane is shaded,

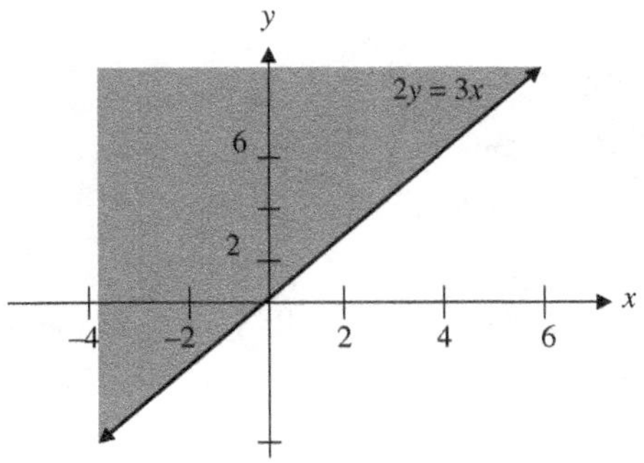

11. For $2x + y \leq 8$, the line $2x + y = 8$ is solid and the appropriate half-plane shaded.

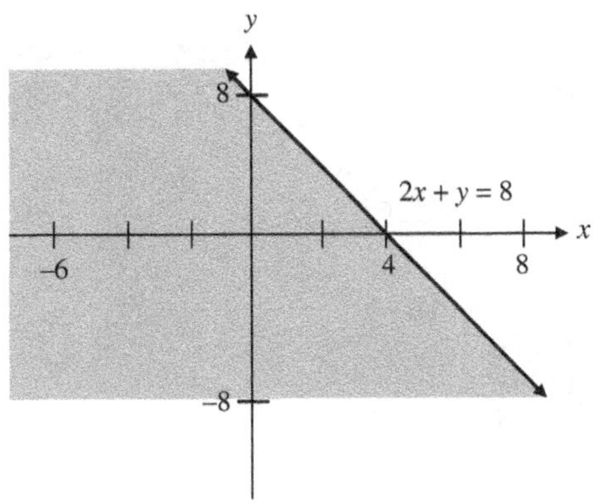

13. For $2x + 5y \geq 10$, the line $2x + 5y = 10$ is solid and the appropriate half-plane shaded.

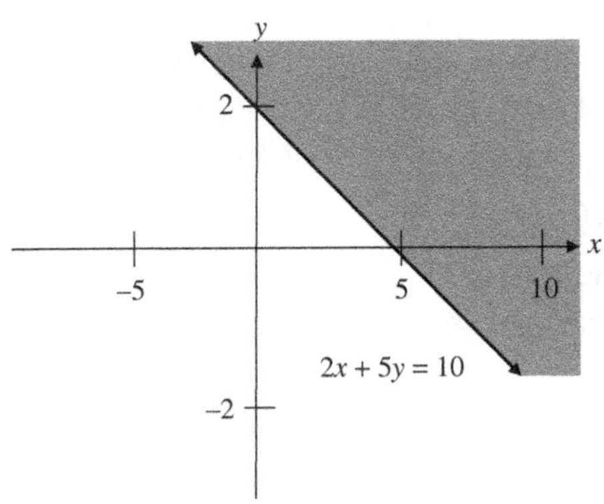

15. For the inequality $y \geq 2x + 3$, the line $y = 2x + 3$ is solid and the appropriate half-plane shaded.

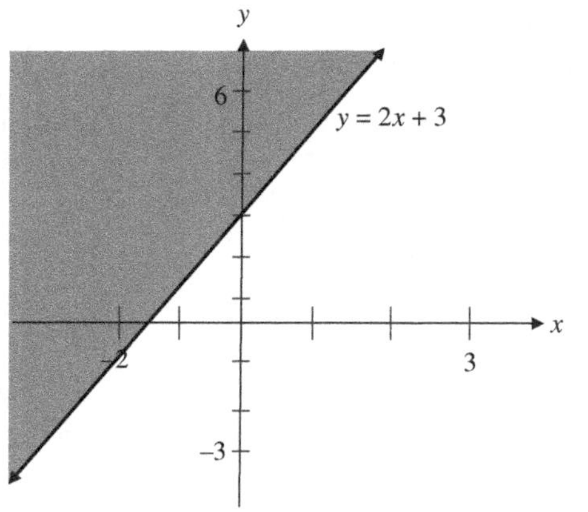

17. For $(1/3)x + (1/2)y \leq 2$, the line $(1/3)x + (1/2)y = 2$ is solid and the appropriate half-plane shaded.

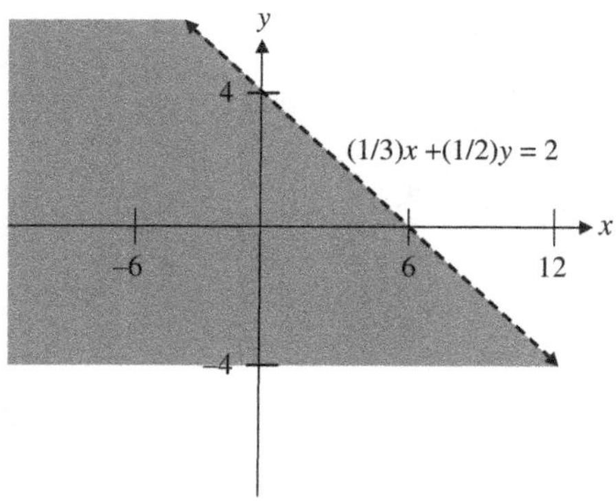

19. The x -intercept is 3 and the y -intercept is 4. Using (3, 0) and (0, 4), the slope is $-4/3$. The line is, $y = (-4/3)x + 4$ and is dashed, as for a strict inequality. A test point yields $y < (-4/3)x + 4$, as the inequality shown.

21. The inequality is $x + y \leq 10$. The line $x + y = 10$ is drawn solid and the appropriate half-plane shaded.

⟶

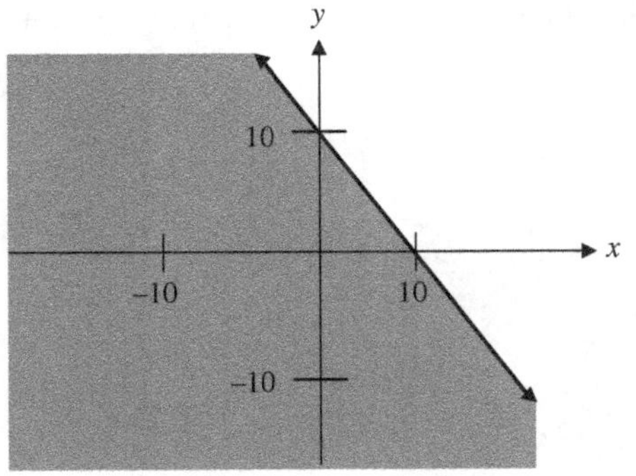

23. Let x = amount of Food A and let y = amount of Food B. The restriction on food intake is $5x + 4y \geq 40$. The combinations are in the first quadrant.

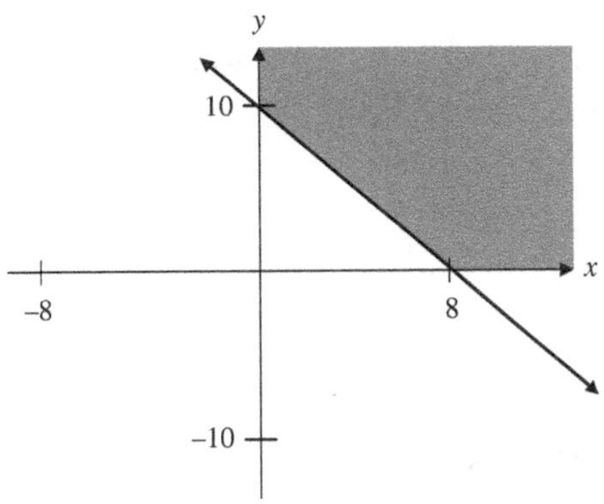

EXERCISES 4.2

1. The corner point is where $x = 1$ and $y = 3$ or (1, 3).

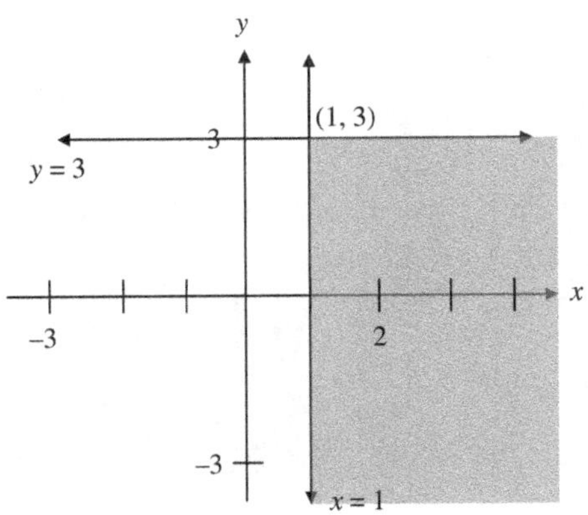

3. Here, the feasible region is a rectangle with vertices (–2, 4), (–2, 1), (5, 4), and (5, 1).

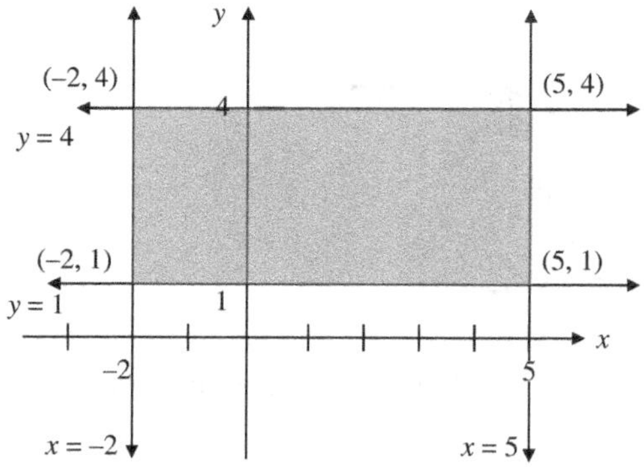

5. The corner point occurs where the two lines intersect. Using elimination or substitution, this point is (5/3, 7/3).

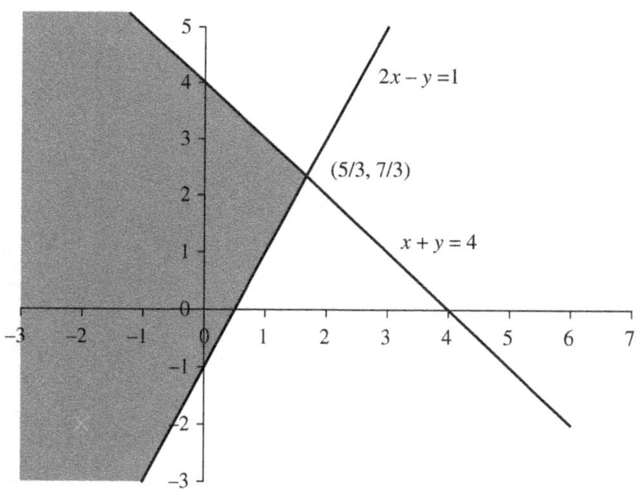

7. The corner point, where $2x + y = 7$ intersects with $x + y = 5$, is (2, 3). The other three corner points are intercepts.

⟶

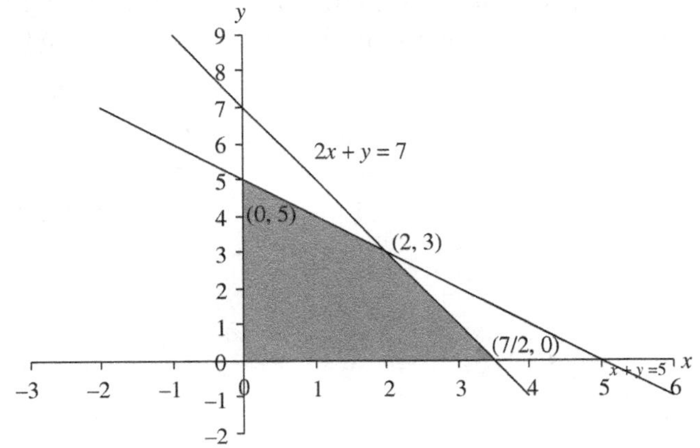

9. The corner point is the origin, as the feasible region is unbounded.

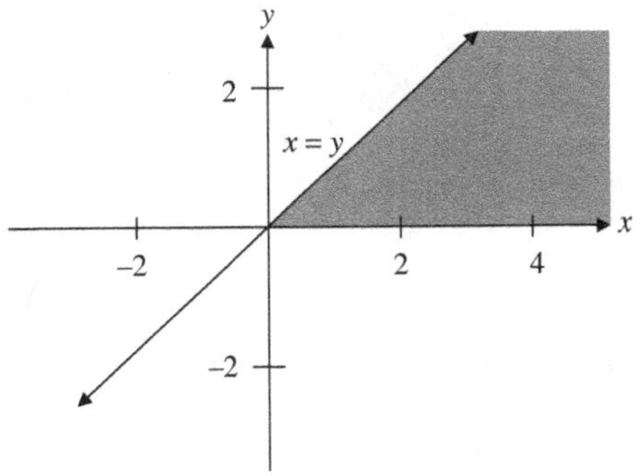

11. The feasible region is the quadrilateral with vertices (2, 0), (2, 3), (4, −3), and (6, −3).

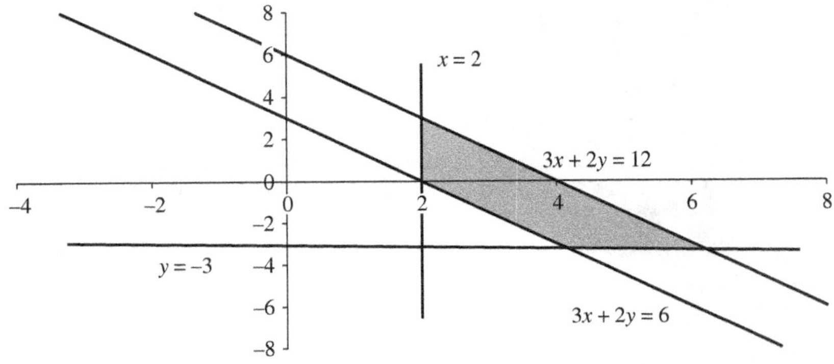

13. The corner points of the feasible region are (0, 0), (0, 4), (3, 1), and (7/2, 0). The interior corner point is where $2x + y = 7$ and $x + y = 4$ intersect.

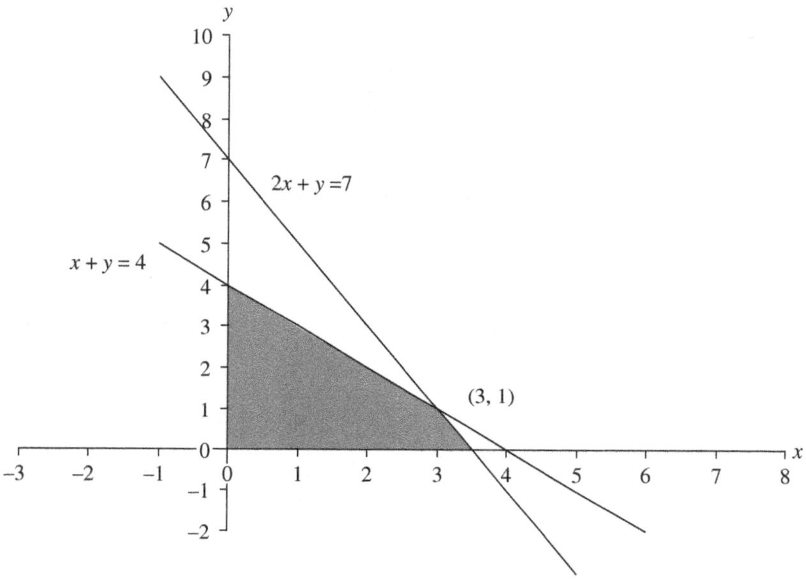

15. The feasible region is unbounded. The only corner point is (2, 3).

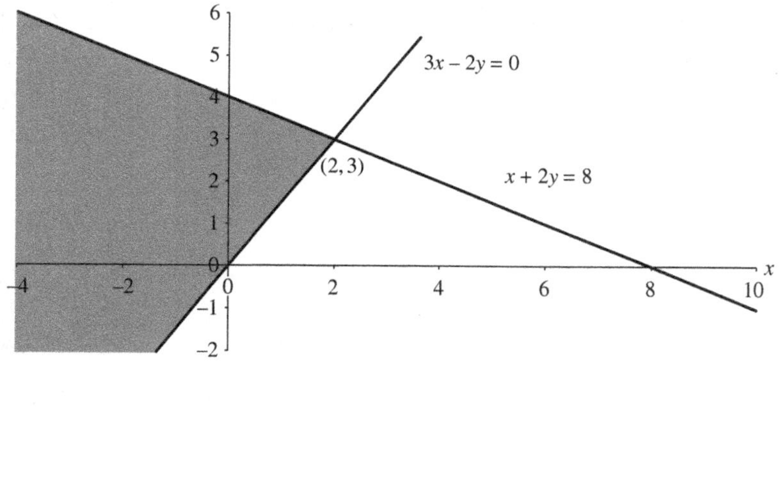

17. The feasible region is a quadrilateral with vertices (0, 6), (0, 10), (10, 0), and (6, 0).

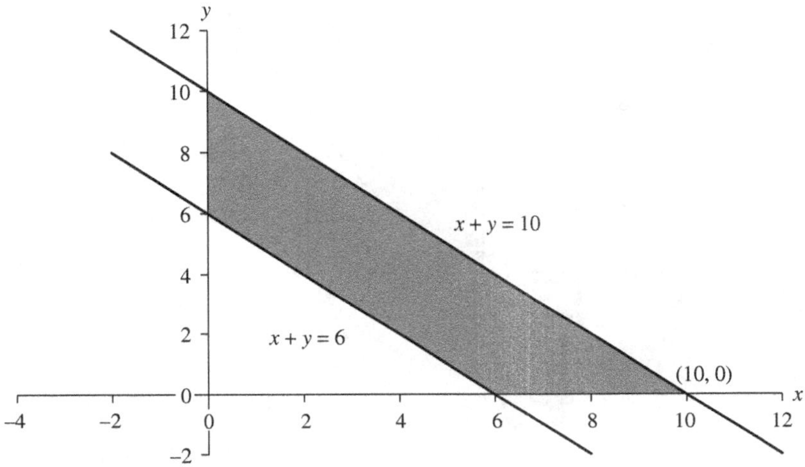

19. The feasible region is the quadrilateral with vertices (0, 0), (0, 20), (10, 15), and (20, 0). The corner point (10, 15) in the interior of quadrant I is where the lines $3x + 2y = 60$ and $x + 2y = 40$ intersect.

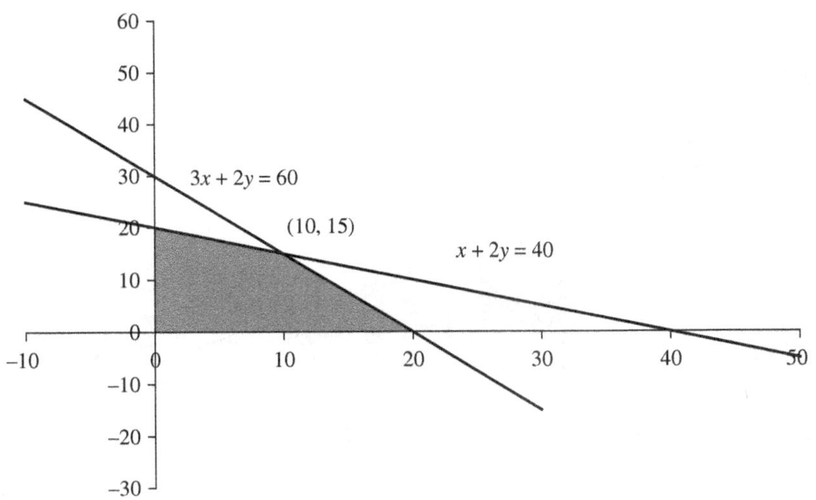

21. The feasible region is a quadrilateral with vertices (0, 0), (0, 5), (1, 0), and (7/3, 8/3).

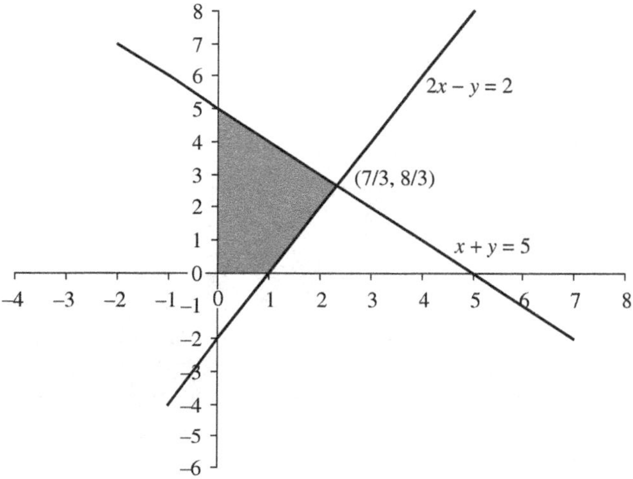

23. The feasible region has five corner points. They are (0, 0), (0, 5), (6, 2), (4, 3), and (7, 0).

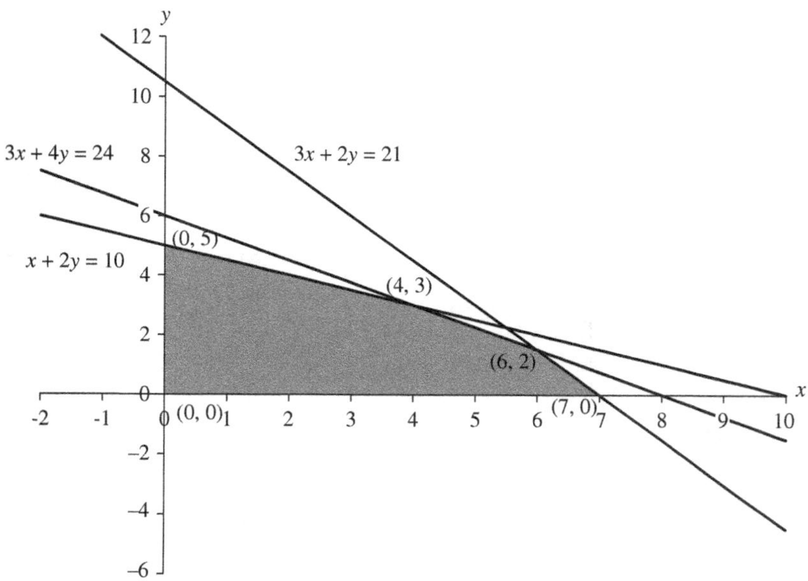

25. The feasible region has five corner points. They are (0, 0), (0, 1), (3, 4), (6, 3), and (5, 0).

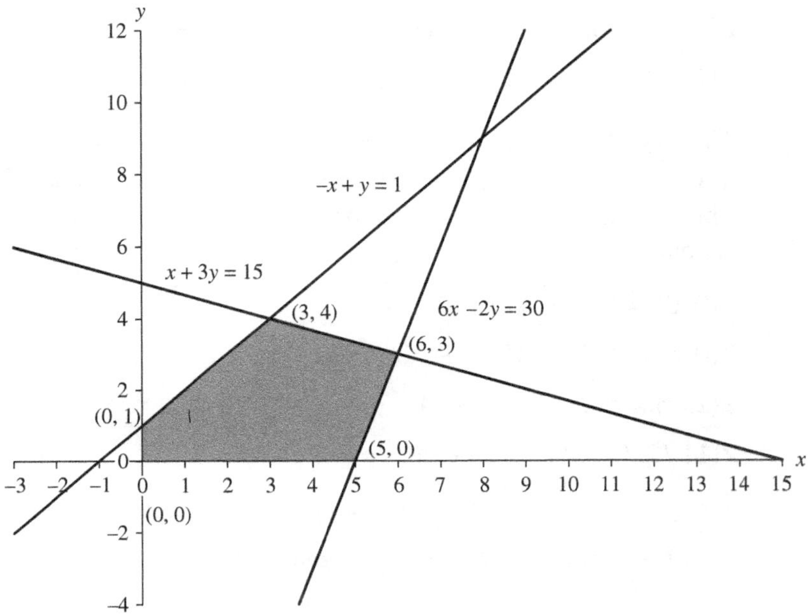

EXERCISES 4.3

For problems 1–5, there are four corner points: (0, 0) (0, 5), (2, 7), and (11/2, 0).

1. Maximize $2x + 3y$ subject to the graphical constraints shown. The maximum must occur at one of the corner points since the feasible region is bounded.

$$Z(0, 0) = 2(0) + 3(0) = 0$$

$$Z(0, 5) = 2(0) + 3(5) = 15$$

$$Z(2, 7) = 2(2) + 3(7) = 25$$

$$Z(11/2, 0) = 2(11/2) + 3(0) = 11$$

The largest value for Z is 25 when $x = 2$ and $y = 7$

3. Maximize $10x + 4y$ subject to the graphical constraints shown. The maximum is at one of the corner points since the feasible region is bounded.

⟶

$$Z(0,0) = 10(0) + 4(0) = 0$$

$$Z(0,5) = 10(0) + 4(5) = 20$$

$$Z(2,7) = 10(2) + 4(7) = 48$$

$$Z(11/2,0) = 10(11/2) + 4(0) = 55$$

The largest value for Z is 55 when $x = 11/2$ and $y = 0$.

5. Maximize $5x - 2y$ subject to the graphical constraints shown.
The maximum is at one of the corner points since the feasible region is bounded.

$$Z(0, 0) = 5(0) - 2(0) = 0$$

$$Z(0, 5) = 5(0) - 2(5) = -10$$

$$Z(2, 7) = 5(2) - 2(7) = -4$$

$$Z(11/2, 0) = 5(11/2) - 2(0) = 55/2$$

The largest value for Z is 55/2 when $x = 11/2$ and $y = 0$.

For problems 6–10, the corner points are (0, 3) (0, 7), (3, 0), and (7, 0).

7. Minimize $x + 3y$ subject to the graphical constraints shown. The minimum is at one of the corner points since the feasible region is bounded.

$$Z(0, 3) = (0) + 3(3) = 9$$

$$Z(0, 7) = (0) + 3(7) = 21$$

$$Z(3, 0) = (3) + 3(0) = 3$$

$$Z(7, 0) = (7) + 3(0) = 7$$

The minimum value for Z is 3 when $x = 3$ and $y = 0$.

9. Minimize $2x + 5y$ subject to the graphical constraints shown. The minimum is at one of the corner points since the feasible region is bounded.

$$Z(0, 3) = 2(0) + 5(3) = 15$$

$$Z(0, 7) = 2(0) + 5(7) = 35$$

$$Z(3, 0) = 2(3) + 5(0) = 6$$

$$Z(7, 0) = 2(7) + 5(0) = 14$$

The minimum value for Z is 6 when $x = 3$ and $y = 0$.

11. The feasible region is a quadrilateral with vertices (0, 0), (0, 7/2), (3, 2), and (4, 0) shown.

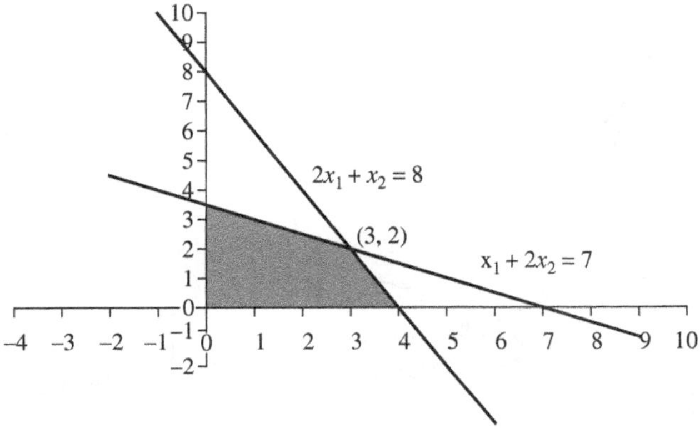

The objective function is evaluated at each corner point

$Z(0, 0) = 2(0) + 3(0) = 0$ $\quad$ $Z(3, 2) = 2(3) + 3(2) = 12$

$Z(0, 7/2) = 2(0) + 3(7/2) = 21/2$ $\quad$ $Z(4, 0) = 2(4) + 3(0) = 8$

The maximum value is 12 when $x = 3$ and $y = 2$.

13. The feasible region is a quadrilateral with vertices (0, 0), (0, 7/2), (25/9, 19/9), and (16/5, 0)

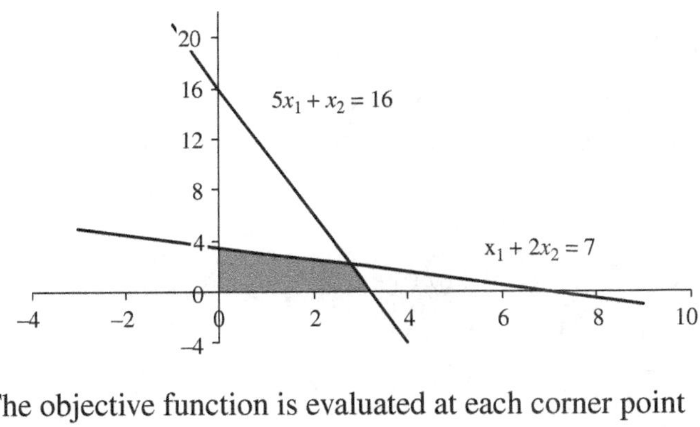

The objective function is evaluated at each corner point

$Z(0, 0) = 5(0) + 4(0) = 0$

$Z(25/9, 19/9) = 5(25/9) + 4(19/9) = 201/9^*$

$Z(0, 7/2) = 5(0) + 4(7/2) = 14$

$Z(16/5, 0) = 5(16/5) + 4(0) = 16$

The maximum value is 201/9 when $x = 25/9$ and $y = 19/9$.

15. The feasible region has five corner points. The corner points are (0, 0), (0, 2), (2, 4), (4, 3), and (5, 0)

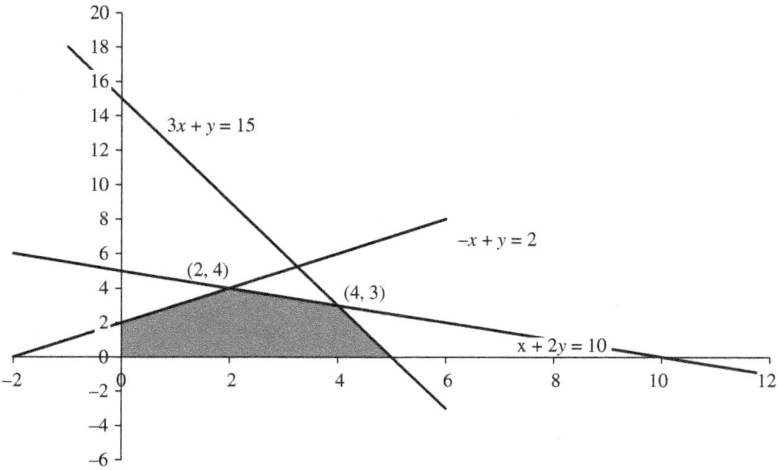

The objective function is evaluated at each corner point

$Z(0, 0) = 5(0) + 2(0) = 0$ $\quad Z(0, 2) = 5(0) + 2(2) = 4$
$Z(2, 4) = 5(2) + 2(4) = 18$ $\quad Z(4, 3) = 5(4) + 2(3) = 26^*$
$Z(5, 0) = 5(5) + 2(0) = 25$

The maximum is 26 when $x = 4$ and $y = 3$.

17. Let $x = \#$ items of x to produce, $y = \#$ items of y to produce. 15 minutes is converted to 1/4 h and 10 min to 1/6 h.

$$\text{Maximize}: Z = 16x + 20y$$

$$\begin{aligned} \text{Subject to}: \quad (1/4)x + (1/6)y &\le 40 \\ 3x + \quad 4y &\le 600 \\ x, y &\ge 0 \end{aligned}$$

19. Let $x = \#$ type A trucks and $y = \#$ type B trucks

$$\text{Minimize}: \quad Z = 3x + 4y$$

$$\begin{aligned} \text{Subject to}: \quad 100x + 150y &\ge 4500 \\ 200x + 150y &\ge 6000 \\ x, y &\ge 0 \end{aligned}$$

SUPPLEMENTARY EXERCISES CHAPTER 4

1. For $x + 2y < 20$, the line $x + 2y = 20$ is drawn dashed and the appropriate half-plane shaded.

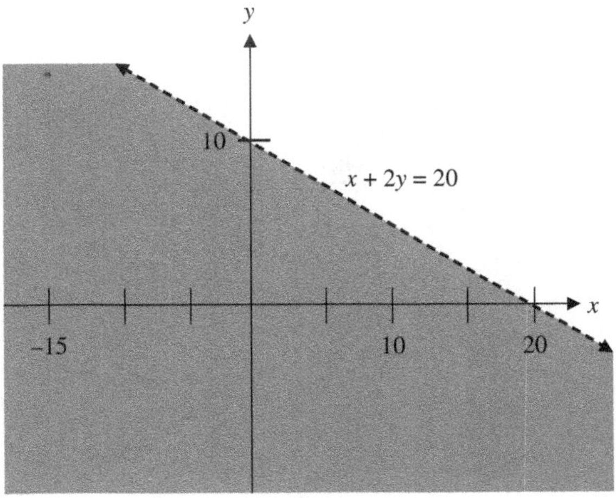

3. For $40x + 30y \leq 120$, the line $40x + 30y = 120$ is solid and the appropriate half-plane shaded.

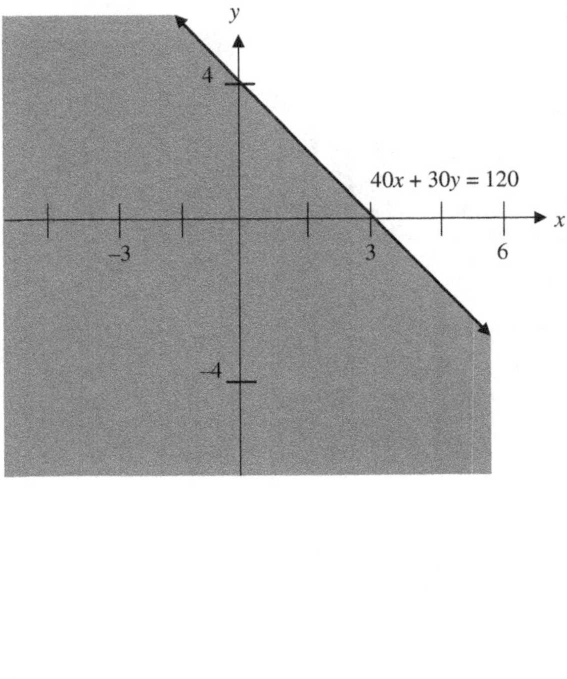

5. The feasible region is a triangle with vertices (0, 0), (0, 30), and (30, 0)

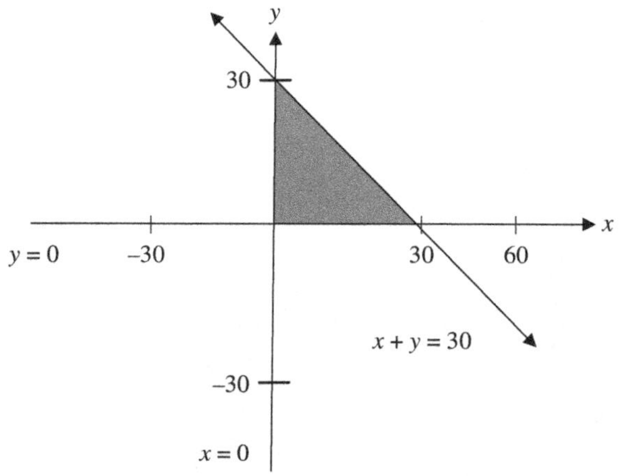

7. The feasible region is a triangle with vertices (0, 0), (0, 10), and (20, 0).

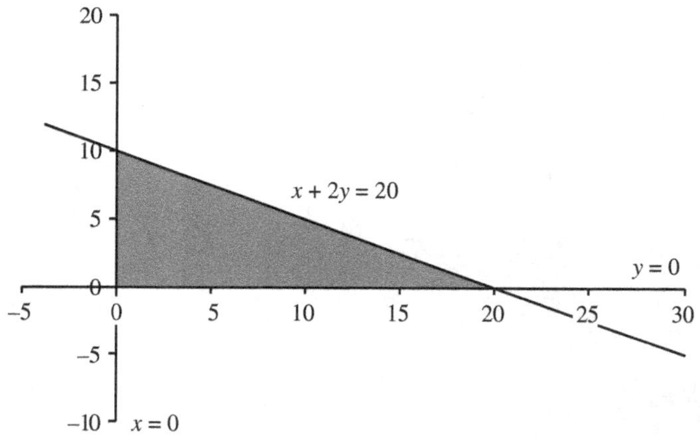

9. The feasible region is a quadrilateral with vertices (0, 0), (0, 60), (20, 50), and (15/2, 0).

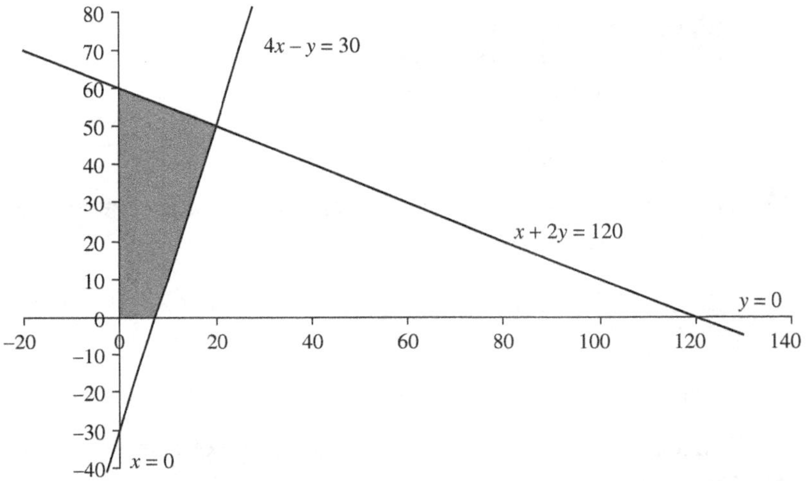

11. The feasible region is a quadrilateral with vertices (0, 0), (40, 20), (100/3, 0), and (0, 100).

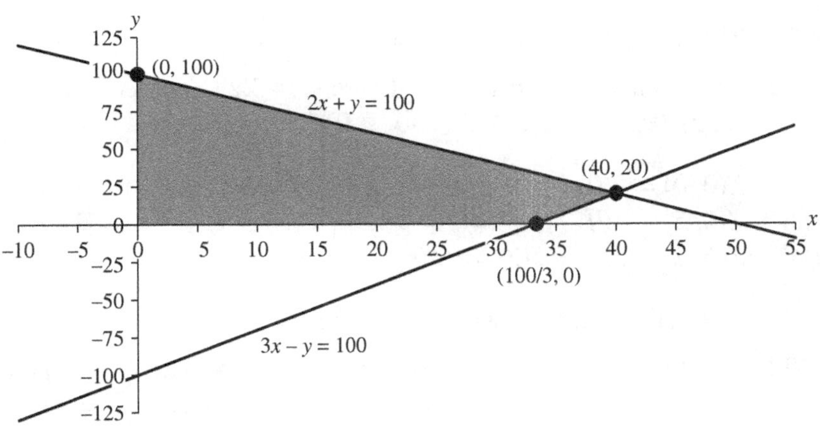

13. The feasible region is a triangle with vertices (0, 0), (0, 5), and (10, 0). The objective function is to maximize $Z = 5x + 2y$. Then,

$$Z(0, 0) = 5(0) + 2(0) = 0$$
$$Z(0, 5) = 5(0) + 2(5) = 10$$
$$Z(10, 0) = 5(10) + 2(0) = 50 \text{(maximum)}$$

The maximum value is 50 when $x = 10$ and $y = 0$.

15. The feasible region is a triangle with vertices (0, 0), (0, 15), and (10, 0). The objective is to maximize $Z = 3x + 5y$. Then,

$$Z(0, 0) = 3(0) + 5(0) = 0$$
$$Z(0, 15) = 3(0) + 5(15) = 75$$
$$Z(10, 0) = 3(10) + 5(0) = 30 \text{(maximum)}$$

The maximum is 75 when $x = 0$ and $y = 15$.

17. The corner points of the feasible region form a quadrilateral with vertices (0, 0), (0, 6), (2, 0), and (4, 2).

$$Z(0, 0) = 20(0) + 10(0) = 0 \qquad Z(0, 6) = 20(0) + 10(6) = 60$$
$$Z(2, 0) = 20(2) + 10(0) = 40 \qquad Z(4, 2) = 20(4) + 10(2) = 100$$

The maximum is 100 when $x = 4$ and $y = 2$.

19. The feasible region is from Supplementary Exercise 18. The objective has changed to $Z = 3x + 2y$

$$Z(0, 0) = 3(0) + 2(0) = 0 \qquad Z(0, 35/2) = 3(0) + 2(35/2) = 35$$
$$Z(20, 0) = 3(20) + 2(0) = 60^{*} \qquad Z(15, 10) = 3(15) + 2(10) = 55$$

The maximum is 60 when $x = 20$ and $y = 0$.

21. The feasible region is from Exercise 4.2.23. The corner points are (0, 0), (0, 5), (6, 2), (4, 3), and (7, 0). The objective is $Z = 3x + 5y$, so

$$Z(0, 0) = 3(0) + 5(0) = 0 \qquad Z(4, 3) = 3(4) + 5(3) = 27$$
$$Z(0, 5) = 3(0) + 5(5) = 25 \qquad Z(7, 0) = 3(7) + 5(0) = 21$$
$$Z(6, 2) = 3(6) + 5(2) = 28^{*}$$

The maximum is 28 when $x = 6$ and $y = 2$.

23. The feasible region is from Exercise 4.2.25. The corner points are (0, 0), (0, 1), (3, 4), (6, 3), and (5, 0). The objective is $Z = 3x - 2y$, so

$$Z(0, 0) = 3(0) - 2(0) = 0 \qquad Z(6, 3) = 3(6) - 2(3) = 12$$
$$Z(0, 1) = 3(0) - 2(1) = -2 \qquad Z(5, 0) = 3(5) - 2(0) = 15^{*}$$
$$Z(3, 4) = 3(3) - 2(4) = 1$$

The maximum is 15 when $x = 5$ and $y = 0$.

CHAPTER 5

LINEAR PROGRAMMING – SIMPLEX METHOD

EXERCISES 5.1

1. It is a Standard Maximization Problem (SMP).

3. It is an SMP.

5. $[1 \quad 1]$, $[0 \quad 5]$, and $[3 \quad 3]$ are feasible. The vector $[5 \quad 2]$ fails the second constraint.

7. $[1 \quad 1 \quad 5]$ and $[0 \quad 6 \quad 0]$ are feasible. $[2 \quad 4 \quad 3]$ and $[4 \quad 3 \quad 5]$ fail the first constraint. There is no need to continue.

9. $[4 \quad 3 \quad 5]$ has all nonnegative values.

$$\begin{aligned} 4 + 3 + 2(5) &\leq 20 \text{ is true} \\ 2(4) + 3 + 2(5) &\leq 25 \text{ is true} \\ 5(4) + 3 + \ \ 5 &\leq 30 \text{ is true.} \end{aligned}$$

Therefore, $[4 \quad 3 \quad 5]$ is feasible.

⟶

Solutions Manual to Accompany Finite Mathematics: Models and Applications, First Edition.
Carla C. Morris and Robert M. Stark.

$\begin{bmatrix}3 & 0 & 7\end{bmatrix}$ has all nonnegative values.

$$\begin{aligned} 3+0+2(7) &\le 20 \text{ is true} \\ 2(3)+0+2(7) &\le 25 \text{ is true} \\ 5(4)+0+7 &\le 30 \text{ is true.} \end{aligned}$$

Therefore, $\begin{bmatrix}3 & 0 & 7\end{bmatrix}$ is feasible.

11. Maximize: $Z = x_1 + x_2$

Subject to: $2x_1 + 3x_2 + s_1 = 9$

$x_1 + x_2 + s_2 = 3$

$x_1, x_2, s_1, s_2 \ge 0.$

13. Maximize: $P = 3x + 4y + 2z$

Subject to: $5x + y + z + s_1 = 1500$

$4x + 2y + z + s_2 = 2000$

$2x + 2y + z + s_3 = 2500$

$x, y, z, s_1, s_2, s_3 \ge 0.$

15. Maximize: $Z = 16x_1 + 20x_2$

Subject to: $(1/4)x_1 + (1/6)x_2 + s_1 = 40$

$3x_1 + 4x_2 + s_2 = 600$

$x_1, x_2, s_1, s_2 \ge 0.$

EXERCISES 5.2

1. The tableau is:

Row	x_1	x_2	s_1	s_2	RHS
1	1	3	1	0	10
2	2	1	0	1	8
Objective	–5	–2	0	0	0

3. The tableau is:

Row	x	y	z	s_1	s_2	s_3	RHS
1	1	2	1	1	0	0	12
2	2	1	3	0	1	0	18
3	1	3	1	0	0	1	20
Objective	−5	−1	−1	0	0	0	0

5. Maximize: $Z = 4x + 3y$

Subject to: $30x + 20y \leq 500$

$5x + 10y \leq 150$

$x, y \geq 0.$

7. Maximize: $Z = 2x_1 + 3x_2$

Subject to: $3x_1 + 5x_2 + 2x_3 \leq 19$

$5x_1 + 2x_2 + 4x_3 \leq 21$

$x_1 + 7x_2 + 3x_3 \leq 22$

$x_1, x_2, x_3 \geq 0.$

9. a) The pivot is the x_1 column since the most negative number in the objective row is −5. The ratios are $10/1 = 10$ and $8/2 = 4$, the minimum is 4, so the pivot element is the "2" in the second row.

b) The pivot is the y column since the −5 is the most negative number in the objective row. The minimum ratio among 19/2, 26/4, and 6/1 is 6/1. The "1" in the third row is the pivot element.

c) The pivot is the x column since the −5 is the most negative number in the objective row. The minimum ratio among 12/1, 18/2, and 20/1 is 18/2. The "2" in the second row is the pivot element.

d) The pivot is the x column since the −3 is the most negative number in the objective row. The minimum ratio among 20/2 and 14/1 is 20/2. The "2" in the first row is the pivot element.

11. The tableau with pivot element in bold is:

Row	x_1	x_2	s_1	s_2	RHS
1	1	3	1	0	10
2	**2**	1	0	1	8
Objective	−5	−2	0	0	0

The pivot is:

Row	x_1	x_2	s_1	s_2	RHS
1	1	3	1	0	10
2	1	1/2	0	1/2	4
Objective	−5	−2	0	0	0

Row	x_1	x_2	s_1	s_2	RHS
1	0	5/2	1	– 1/2	6
2	1	1/2	0	1/2	4
Objective	0	1/2	0	5/2	20

The pivot completed, the current solution is optimal since there are no negative numbers in the objective row. The current (optimal) solution is $x_1^* = 4$, $x_2^* = 0$, $s_1 = 6$, $s_2 = 0$, and the objective value is 20. It is useful to recheck the objective and each inequality.

13. The tableau, pivot element in bold, is:

Row	x	y	z	s_1	s_2	s_3	RHS
1	1	2	1	1	0	0	12
2	**2**	1	3	0	1	0	18
3	1	3	1	0	0	1	20
Objective	−5	−1	−1	0	0	0	0

⟶

The pivot is:

Row	x	y	z	s_1	s_2	s_3	RHS
1	0	3/2	– 1/2	1	– 1/2	0	3
2	**1**	1/2	3/2	0	1/2	0	9
3	0	5/2	– 1/2	0	– 1/2	1	11
Objective	0	1/2	13/2	0	5/2	0	45

The current (optimal) solution is $x^* = 9, y^* = 0, z^* = 0, s_1 = 3, s_2 = 0, s_3 = 11$, and the objective value is 45.

15. The tableau, pivot element in bold, is:

Row	x	y	s_1	s_2	RHS
1	**30**	20	1	0	500
2	5	10	0	1	150
Objective	–4	–3	0	0	0

The pivot is:

Row	x	y	s_1	s_2	RHS
1	1	2/3	1/30	0	50/3
2	0	20/3	$-1/6$	1	200/3
Objective	0	$-1/3$	2/15	0	200/3

The current (nonoptimal) solution is $x = 50/3$, $y = 0$, $s_1 = 0$, $s_2 = 200/3$, and the objective value is 200/3.

17. The tableau with pivot element in bold is:

Row	x_1	x_2	x_3	s_1	s_2	s_3	RHS
1	3	5	2	1	0	0	19
2	5	2	4	0	1	0	21
3	1	**7**	3	0	0	1	22
Objective	–2	–3	0	0	0	0	0

→

The pivot is:

Row	x_1	x_2	x_3	s_1	s_2	s_3	RHS
1	3	5	2	1	0	0	19
2	5	2	4	0	1	0	21
3	1/7	1	3/7	0	0	1/7	22/7
Objective	−2	−3	0	0	0	0	0

Row	x_1	x_2	x_3	s_1	s_2	s_3	RHS
1	16/7	0	$-1/7$	1	0	$-5/7$	23/7
2	33/7	0	22/7	0	1	$-2/7$	103/7
3	1/7	1	3/7	0	0	1/7	22/7
Objective	−11/7	0	9/7	0	0	3/7	66/7

The current (nonoptimal) solution is $x_1 = 0$, $x_2 = 22/7$, $x_3 = 0$, $s_1 = 23/7$, $s_2 = 103/7$, $s_3 = 0$, and the objective value is 66/7.

19. The current solution is $x_1 = 0$, $x_2 = 0$, $x_3 = 20$, $x_4 = 75$, $s_1 = 15$, $s_2 = 10$, $s_3 = 0$, $s_4 = 0$, and the objective value is 230. It is not optimal as the negative in the objective row indicates that further pivoting is necessary.

EXERCISES 5.3

1. The tableaus for the Simplex Method are:

Row	x_1	x_2	s_1	s_2	RHS
1	2	1	1	0	8
2	1	**2**	0	1	7
Objective	−2	−3	0	0	0

⟶

Row	x_1	x_2	s_1	s_2	RHS
1	2	1	1	0	8
2	1/2	1	0	1/2	7/2
Objective	−2	−3	0	0	0

Row	x_1	x_2	s_1	s_2	RHS
1	**3/2**	0	1	−1/2	9/2
2	1/2	1	0	1/2	7/2
Objective	−1/2	0	0	3/2	21/2

Row	x_1	x_2	s_1	s_2	RHS
1	1	0	2/3	−1/3	3
2	1/2	1	0	1/2	7/2
Objective	−1/2	0	0	3/2	21/2

Row	x_1	x_2	s_1	s_2	RHS
1	1	0	2/3	−1/3	3
2	0	1	−1/3	2/3	2
Objective	0	0	1/3	4/3	12

The optimal solution is $x_1^* = 3$, $x_2^* = 2$, $s_1 = 0$, $s_2 = 0$, and the objective value is 12.

3. The tableaus for the Simplex Method are:

Row	x_1	x_2	s_1	s_2	RHS
1	1	1	1	0	24
2	**2**	1	0	1	32
Objective	−3	−2	0	0	0

⟶

Row	x_1	x_2	s_1	s_2	RHS
1	1	1	1	0	24
2	1	1/2	0	1/2	16
Objective	−3	−2	0	0	0

Row	x_1	x_2	s_1	s_2	RHS
1	0	***1/2***	1	−1/2	8
2	1	1/2	0	1/2	16
Objective	0	−1/2	0	3/2	48

Row	x_1	x_2	s_1	s_2	RHS
1	0	1	2	−1	16
2	1	1/2	0	1/2	16
Objective	0	−1/2	0	3/2	48

Row	x_1	x_2	s_1	s_2	RHS
1	0	1	2	−1	16
2	1	0	−1	1	8
Objective	0	0	1	1	56

The optimal solution is $x_1^* = 8$, $x_2^* = 16$, $s_1 = 0$, $s_2 = 0$, and the objective value is 56.

5. The tableaus for the Simplex Method are:

Row	x_1	x_2	s_1	s_2	RHS
1	1	***1***	1	0	10
2	2	1	0	1	14
Objective	−1	−3	0	0	0

⟶

Row	x_1	x_2	s_1	s_2	RHS
1	1	1	1	0	10
2	1	0	−1	1	4
Objective	2	0	3	0	30

The optimal solution is $x_1^* = 0$, $x_2^* = 10$, $s_1 = 0$, $s_2 = 4$, and the objective value is 30.

7. The tableaus for the Simplex Method are:

Row	x_1	x_2	x_3	s_1	s_2	s_3	RHS
1	2	1	1	1	0	0	50
2	*1*	3	0	0	1	0	20
3	0	1	1	0	0	1	15
Objective	−5	−1	−1	0	0	0	0

Row	x_1	x_2	x_3	s_1	s_2	s_3	RHS
1	0	−5	*1*	1	−2	0	10
2	1	3	0	0	1	0	20
3	0	1	1	0	0	1	15
Objective	0	14	−1	0	5	0	100

Row	x_1	x_2	x_3	s_1	s_2	s_3	RHS
1	0	−5	1	1	−2	0	10
2	1	3	0	0	1	0	20
3	0	6	0	−1	2	1	5
Objective	0	9	0	1	3	0	110

The optimal solution is $x_1^* = 20$, $x_2^* = 0$, $x_3^* = 10$, $s_1 = 0$, $s_2 = 0$, $s_3 = 5$, and the objective value is 110.

9. The tableaus for the Simplex Method are:

Row	x_1	x_2	x_3	s_1	s_2	s_3	RHS
1	2	1	1	1	0	0	4
2	1	2	0	0	1	0	10
3	0	1	***3***	0	0	1	8
Objective	−2	−3	−4	0	0	0	0

Row	x_1	x_2	x_3	s_1	s_2	s_3	RHS
1	2	1	1	1	0	0	4
2	1	2	0	0	1	0	10
3	0	1/3	1	0	0	1/3	8/3
Objective	−2	−3	−4	0	0	0	0

Row	x_1	x_2	x_3	s_1	s_2	s_3	RHS
1	***2***	2/3	0	1	0	−1/3	4/3
2	1	2	0	0	1	0	10
3	0	1/3	1	0	0	1/3	8/3
Objective	−2	−5/3	0	0	0	4/3	32/3

Another pivot is indicated.

Row	x_1	x_2	x_3	s_1	s_2	s_3	RHS
1	1	1/3	0	1/2	0	−1/6	2/3
2	1	2	0	0	1	0	10
3	0	1/3	1	0	0	1/3	8/3
Objective	−2	−5/3	0	0	0	4/3	32/3

⟶

Row	x_1	x_2	x_3	s_1	s_2	s_3	RHS
1	1	***1/3***	0	1/2	0	−1/6	2/3
2	0	5/3	0	−1/2	1	1/6	28/3
3	0	1/3	1	0	0	1/3	8/3
Objective	0	−1	0	1	0	1	12

Another pivot is indicated.

Row	x_1	x_2	x_3	s_1	s_2	s_3	RHS
1	3	1	0	3/2	0	−1/2	2
2	0	5/3	0	−1/2	1	1/6	28/3
3	0	1/3	1	0	0	1/3	8/3
Objective	0	−1	0	1	0	1	12

Row	x_1	x_2	x_3	s_1	s_2	s_3	RHS
1	3	1	0	3/2	0	−1/2	2
2	−5	0	0	−3	1	1	6
3	−1	0	1	−1/2	0	1/2	2
Objective	3	0	0	5/2	0	1/2	14

The optimal solution is $x_1^* = 0, x_2^* = 2, x_3^* = 2, s_1 = 0, s_2 = 6, s_3 = 0$, and the objective value is 14.

11. There is a tie in the minimum ratio, so either the "4"or the "1" in the x_2 column can be used as the pivot element. Here, the "1" is used.

Row	x_1	x_2	s_1	s_2	s_3	RHS
1	50	75	1	0	0	20,000
2	3	4	0	1	0	800
3	0	*1*	0	0	1	200
Objective	−25	−35	0	0	0	0

⟶

Row	x_1	x_2	s_1	s_2	s_3	RHS
1	50	0	1	0	−75	5000
2	*3*	0	0	1	−4	0
3	0	1	0	0	1	200
Objective	−25	0	0	0	35	7000

Row	x_1	x_2	s_1	s_2	s_3	RHS
1	50	0	1	0	−75	5000
2	*1*	0	0	1/3	−4/3	0
3	0	1	0	0	1	200
Objective	−25	0	0	0	35	7000

Row	x_1	x_2	s_1	s_2	s_3	RHS
1	0	0	1	−50/3	−25/3	5000
2	1	0	0	1/3	−4/3	0
3	0	1	0	0	1	200
Objective	0	0	0	25/3	5/3	7000

The optimal solution is $x_1^* = 0$, $x_2^* = 200$, $s_1 = 5000$, $s_2 = 0$, and $s_3 = 0$. The objective value is 7000.

13. The SMP in matrix format is:

$$\begin{aligned} \text{Maximize:} \quad & Z = \mathbf{C}^{\mathrm{T}}\mathbf{X} \\ \text{Subject to:} \quad & \mathbf{AX} \leq \mathbf{B} \\ & \mathbf{X} \geq 0, \end{aligned}$$

where $\mathbf{A} = \begin{bmatrix} 1 & 3 & 0 \\ 2 & 1 & 1 \\ 1 & 2 & 4 \end{bmatrix}$; $\mathbf{B} = \begin{bmatrix} 30 \\ 60 \\ 125 \end{bmatrix}$; $\mathbf{C} = \begin{bmatrix} 5 \\ 3 \\ 2 \end{bmatrix}$; $\mathbf{X} = \begin{bmatrix} x_1 \\ x_2 \\ x_3 \end{bmatrix}$.

EXERCISES 5.4

1. The dual is

$$\begin{aligned} \text{Minimize:}\quad & V = 12y_1 + 18y_2 + 20y_3 \\ \text{Subject to:}\quad & y_1 + 2y_2 + y_3 \geq 5 \\ & 2y_1 + y_2 + 3y_3 \geq 1 \\ & 3y_1 + 3y_2 + y_3 \geq 1 \\ & y_1,\ y_2,\ y_3 \geq 0. \end{aligned}$$

3. The dual is

$$\begin{aligned} \text{Maximize:}\quad & V = 10y_1 + 15y_2 \\ \text{Subject to:}\quad & y_1 + 3y_2 \leq 11 \\ & 2y_1 + y_2 \leq 7 \\ & y_1,\ y_2 \geq 0. \end{aligned}$$

Row	y_1	y_2	s_1	s_2	RHS
1	1	***3***	1	0	11
2	2	1	0	1	7
Objective	−10	−15	0	0	0

Row	y_1	y_2	s_1	s_2	RHS
1	1/3	1	1/3	0	11/3
2	2	1	0	1	7
Objective	−10	−15	0	0	0

Row	y_1	y_2	s_1	s_2	RHS
1	1/3	1	1/3	0	11/3
2	***5/3***	0	−1/3	1	10/3
Objective	−5	0	5	0	55

⟶

Row	y_1	y_2	s_1	s_2	RHS
1	1/3	1	1/3	0	11/3
2	1	0	−1/5	3/5	2
Objective	−5	0	5	0	55

Row	y_1	y_2	s_1	s_2	RHS
1	0	1	2/5	−1/5	3
2	1	0	−1/5	3/5	2
Objective	0	0	4	3	65

The solution to the primal is in this last objective row. The optimal solution is $x_1 = 4$, $x_2 = 3$, $s_1 = 0$, $s_2 = 0$, and the objective value is 65.

5. The dual is

Maximize: $V = 46y_1 + 40y_2$

Subject to: $3y_1 + 5y_2 \leq 3$

$4y_1 + 3y_2 \leq 2$

$y_1, y_2 \geq 0.$

Row	y_1	y_2	s_1	s_2	RHS
1	3	5	1	0	3
2	***4***	3	0	1	2
Objective	−46	−40	0	0	0

Row	y_1	y_2	s_1	s_2	RHS
1	3	5	1	0	3
2	1	3/4	0	1/4	1/2
Objective	−46	−40	0	0	0

⟶

Row	y_1	y_2	s_1	s_2	RHS
1	0	***11/4***	1	−3/4	3/2
2	1	3/4	0	1/4	1/2
Objective	0	−11/2	0	23/2	23

Row	y_1	y_2	s_1	s_2	RHS
1	0	1	4/11	−3/11	6/11
2	1	3/4	0	1/4	1/2
Objective	0	−11/2	0	23/2	23

Row	y_1	y_2	s_1	s_2	RHS
1	0	1	4/11	−3/11	6/11
2	1	0	−3/11	5/11	1/11
Objective	0	0	2	10	26

The primal solution is in the objective row of the last tableau. The optimal solution is $x_1 = 2$, $x_2 = 10$, $s_1 = 0$, $s_2 = 0$, and the objective value is 26.

7. The dual is

Maximize: $V = 12y_1 + 9y_2 + 15y_3$

Subject to: $4y_1 + y_2 + y_3 \leq 2$

$y_1 + y_2 + y_3 \leq 5$

$y_1, y_2, y_3 \geq 0.$

Row	y_1	y_2	y_3	s_1	s_2	RHS
1	4	1	1	1	0	2
2	1	1	1	0	1	5
Objective	−12	−9	−15	0	0	0

⟶

Row	y_1	y_2	y_3	s_1	s_2	RHS
1	4	1	1	1	0	2
2	−3	0	0	−1	1	3
Objective	48	6	0	15	0	30

The primal solution is in the objective row of the last tableau. The optimal solution is $x_1 = 15$, $x_2 = 0$, $s_1 = 48$, $s_2 = 6$, $s_3 = 0$ and the objective value is 30.

EXERCISES 5.5

1. First, rewrite the Linear Program (LP) as:

$$\text{Maximize:} \quad -Z = -4x_1 - 3x_2$$

$$-2x_1 - x_2 \leq -25$$

$$-x_1 - x_2 \leq -20$$

$$x_1 - 5x_2 \leq 4$$

$$x_1,\ x_2 \geq 0.$$

The initial tableau is:

x_1	x_2	s_1	s_2	s_3	RHS
−2	−1	1	0	0	−25
−1	−1	0	1	0	−20
1	−5	0	0	1	4
4	3	0	0	0	0

Since there are negative numbers in the RHS constraint row, they require removal. First, one removes the −20 by allowing x_1 to enter the basis.

⟶

x_1	x_2	s_1	s_2	s_3	RHS
−2	−1	1	0	0	−25
1	1	0	−1	0	20
1	−5	0	0	1	4
4	3	0	0	0	0

In the first row, its −2 is eliminated and x_2 enters the basis.

x_1	x_2	s_1	s_2	s_3	RHS
0	1	1	−2	0	15
1	1	0	−1	0	20
0	**−6**	0	1	1	−16
0	−1	0	4	0	−80

Another negative arises in the third RHS constraint row as x_2 enters the basis as it has a negative value and is in the same row as the −16.

x_1	x_2	s_1	s_2	s_3	RHS
0	1	1	−2	0	15
1	1	0	−1	0	20
0	**1**	0	−1/6	−1/6	8/3
0	−1	0	4	0	−80

x_1	x_2	s_1	s_2	s_3	RHS
0	0	1	−11/6	**1/6**	37/3
1	0	0	−5/6	1/6	52/3
0	1	0	−1/6	−1/6	8/3
0	0	0	23/6	−1/6	−232/3

The Simplex Method completes the solution.

⟶

x_1	x_2	s_1	s_2	s_3	RHS
0	0	1	$-11/6$	**1/6**	37/3
1	0	0	$-5/6$	1/6	52/3
0	1	0	$-1/6$	$-1/6$	8/3
0	0	0	23/6	$-1/6$	$-232/3$

x_1	x_2	s_1	s_2	s_3	RHS
0	0	6	-11	**1**	74
1	0	0	$-5/6$	1/6	52/3
0	1	0	$-1/6$	$-1/6$	8/3
0	0	0	23/6	$-1/6$	$-232/3$

x_1	x_2	s_1	s_2	s_3	RHS
0	0	6	-11	1	74
1	0	-1	1	0	5
0	1	1	-2	0	15
0	0	1	2	0	-65

The solution is $x_1^* = 5$, $x_2^* = 15$, $s_1 = 0$, $s_2 = 0$, $s_3 = 74$, and the objective value is 65.

3. First, rewrite the LP as:

$$\text{Maximize:} \quad Z = 5x_1 + 6x_2$$
$$4x_1 + 7x_2 \leq 28$$
$$-x_1 + 3x_2 \leq -2$$
$$x_1, \; x_2 \geq 0.$$

⟶

The initial tableau is:

x_1	x_2	s_1	s_2	RHS
4	7	1	0	28
−1	3	0	1	−2
−5	−6	0	0	0

Since there is a negative in the RHS constraint row, it must be made positive by finding another negative number in the same row to allow another variable to enter the basis. In this case, x_1 enters the basis.

x_1	x_2	s_1	s_2	RHS
4	7	1	0	28
1	−3	0	−1	2
−5	−6	0	0	0

x_1	x_2	s_1	s_2	RHS
0	**19**	1	4	20
1	−3	0	−1	2
0	−21	0	−5	10

Now using the Simplex Method:

x_1	x_2	s_1	s_2	RHS
0	1	1/19	4/19	20/19
1	−3	0	−1	2
0	−21	0	−5	10

x_1	x_2	s_1	s_2	RHS
0	1	1/19	**4/19**	20/19
1	0	3/19	−7/19	98/19
0	0	21/19	−11/19	610/19

⟶

x_1	x_2	s_1	s_2	RHS
0	19/4	1/4	1	5
1	0	3/19	$-7/19$	98/19
0	0	21/19	$-11/19$	610/19

x_1	x_2	s_1	s_2	RHS
0	19/4	1/4	1	5
1	7/4	1/4	0	7
0	11/4	5/4	0	35

The optimal solution is $x_1^* = 7$, $x_2^* = 0$, $s_1 = 0$, $s_2 = 5$, and the maximum value is 35.

5. First, rewrite the LP as:

$$\begin{aligned}
\text{Maximize:}\quad & -Z = -x_1 - 2x_2^+ + 2x_2^- + x_3^+ - x_3^- - x_4 \\
\text{Subject to:}\quad & x_1 - x_2^+ + x_2^- + x_3^+ - x_3^- \leq 4 \\
& x_1 + 2x_2^+ - 2x_2^- + 2x_4 \leq 6 \\
& x_2^+ - x_2^- - 2x_3^+ + 2x_3^- + x_4 \leq 2 \\
& -x_1 + 2x_2^+ - 2x_2^- + x_3^+ - x_3^- \leq 2 \\
& x_1,\, x_2^-,\, x_2^+,\, x_3^-,\, x_3^+,\, x_4 \geq 0
\end{aligned}$$

The initial tableau is:

x_1	x_2^+	x_2^-	x_3^+	x_3^-	x_4	s_1	s_2	s_3	s_4	RHS
1	-1	**1**	1	-1	0	1	0	0	0	4
1	1	-1	0	0	2	0	1	0	0	6
0	1	-1	-2	2	1	0	0	1	0	2
-1	2	-2	1	-1	0	0	0	0	1	2
1	2	-2	-1	1	1	0	0	0	0	0

⟶

x_1	x_2^+	x_2^-	x_3^+	x_3^-	x_4	s_1	s_2	s_3	s_4	RHS
1	−1	1	1	−1	0	1	0	0	0	4
2	0	0	1	−1	2	1	1	0	0	10
1	0	0	−1	**1**	1	1	0	1	0	6
1	0	0	3	−3	0	2	0	0	1	10
3	0	0	1	−1	1	2	0	0	0	8

x_1	x_2^+	x_2^-	x_3^+	x_3^-	x_4	s_1	s_2	s_3	s_4	RHS
0	−1	1	0	0	1	2	0	1	0	10
3	0	0	0	0	3	2	1	1	0	16
1	0	0	−1	1	1	1	0	1	0	6
4	0	0	0	0	3	5	0	3	1	28
4	0	0	0	0	2	3	0	1	0	14

The optimal solution is $x_1^* = 0$, $x_2^+ = 0$, $x_2^- = 10$, $(x_2^* = -10)$, $x_3^+ = 0$, $x_3^- = 6$, $(x_3^* = -6)$, $x_4^* = 0$, $s_1 = 0$, $s_2 = 16$, $s_3 = 0$, $s_4 = 28$ and the minimum value is -14.

SUPPLEMENTARY EXERCISES CHAPTER 5

1. The right-hand sides are nonnegative, the inequalities are of the less than or equal to type, and it is a maximization. However, x_4 is unrestricted, so it is not an SMP.

3. Maximize: $Z = 4x_1 + 3x_2$

Subject to:

$$\begin{aligned} 2x_1 + 3x_2 + 5x_3 + s_1 \qquad\qquad &= 23 \\ 3x_1 + x_2 + x_3 \quad + s_2 \qquad &= 8 \\ x_1 + 5x_2 + 3x_3 \qquad\quad + s_3 &= 20 \\ x_1,\ x_2,\ x_3 \geq 0. & \end{aligned}$$

5. The initial tableau is:

x_1	x_2	x_3	s_1	s_2	s_3	RHS
2	5	1	1	0	0	25
3	1	3	0	1	0	21
1	1	0	0	0	1	7
−1	−2	−7	0	0	0	0

7. Maximize: $Z = 4x_1 + x_2 + 3x_3 + 2x_4$

Subject to:
$$2x_1 + x_2 + 3x_3 + 5x_4 \leq 25$$
$$x_1 + 2x_3 + 5x_4 \leq 34$$
$$3x_2 + x_3 + 4x_4 \leq 59$$
$$3x_1 + x_2 + 2x_3 + x_4 \leq 17$$
$$x_1,\ x_2,\ x_3,\ x_4 \geq 0.$$

9. The pivot is the x_1 column since it has the smallest negative number in the objective row. The three ratios are $100/5 = 20$, $75/3 = 25$, and $240/4 = 60$. Therefore, using the minimum ratio, the 5 in the first constraint is the pivot element (in bold italics).

x_1	x_2	x_3	s_1	s_2	s_3	RHS
5	25	10	1	0	0	100
3	15	1	0	1	0	75
4	6	15	0	0	1	240
−4	−3	0	0	0	0	0

11. The pivot is the x_2 column and the pivot element is the 5 in that column.

x_1	x_2	s_1	s_2	RHS
1	5	1	0	200
2	3	0	1	134
−3	−7	0	0	0

⟶

x_1	x_2	s_1	s_2	RHS
1/5	1	1/5	0	40
2	3	0	1	134
−3	−7	0	0	0

Performing the pivot yields the following tableau.

x_1	x_2	s_1	s_2	RHS
1/5	1	1/5	0	40
7/5	0	−3/5	1	14
−8/5	0	7/5	0	280

13. The pivot is the x_1 column and the pivot element is the 5 there.

x_1	x_2	s_1	s_2	RHS
5	1	1	0	16
1	2	0	1	7
−5	−4	0	0	0

x_1	x_2	s_1	s_2	RHS
1	1/5	1/5	0	16/5
1	2	0	1	7
−5	−4	0	0	0

⟶

Performing the pivot yields the following tableau.

x_1	x_2	s_1	s_2	RHS
1	1/5	1/5	0	16/5
0	***9/5***	$-1/5$	1	19/5
0	–3	1	0	16

The tableau indicates that another pivot is necessary.

x_1	x_2	s_1	s_2	RHS
1	1/5	1/5	0	16/5
0	1	$-1/9$	5/9	19/9
0	–3	1	0	16

x_1	x_2	s_1	s_2	RHS
1	0	2/9	$-1/9$	25/9
0	1	$-1/9$	5/9	19/9
0	0	2/3	5/3	67/3

The optimal solution is $x_1 = 25/9$ and $x_2 = 19/9$, and the maximum value of the objective is 67/3.

15. The pivot column is x_2 and the element the 4 in the third row is to become basic.

x_1	x_2	x_3	s_1	s_2	s_3	RHS
4	2	2	1	0	0	40
8	10	5	0	1	0	90
4	***4***	3	0	0	1	30
–3	–5	–4	0	0	0	0

⟶

x_1	x_2	x_3	s_1	s_2	s_3	RHS
4	2	2	1	0	0	40
8	10	5	0	1	0	90
1	1	3/4	0	0	1/4	15/2
−3	−5	−4	0	0	0	0

x_1	x_2	x_3	s_1	s_2	s_3	RHS
2	0	1/2	1	0	−1/2	25
−2	0	−5/2	0	1	−5/2	15
1	1	***3/4***	0	0	1/4	15/2
2	0	−1/4	0	0	5/4	75/2

Another pivot is necessary.

x_1	x_2	x_3	s_1	s_2	s_3	RHS
2	0	1/2	1	0	−1/2	25
−2	0	−5/2	0	1	−5/2	15
4/3	4/3	1	0	0	1/3	10
2	0	−1/4	0	0	5/4	75/2

x_1	x_2	x_3	s_1	s_2	s_3	RHS
4/3	−2/3	0	1	0	−2/3	20
4/3	10/3	0	0	1	−5/3	40
4/3	4/3	1	0	0	1/3	10
7/3	1/3	0	0	0	4/3	40

The optimal solution is $x_1^* = 0, x_2^* = 0$, and $x_3^* = 10$. The optimal value is 40.

17. The dual is

$$\begin{aligned} \text{Maximize:} \quad & V = 48y_1 + 70y_2 + 96y_3 \\ \text{Subject to:} \quad & 3y_1 + 3y_2 + 4y_3 \leq 18 \\ & 2y_1 + 2y_2 + 3y_3 \leq 15 \\ & y_1 + 2y_2 + 2y_3 \leq 10 \\ & y_1, y_2, y_3 \geq 0. \end{aligned}$$

The initial tableau is as follows:

Row	y_1	y_2	y_3	s_1	s_2	s_3	RHS
1	3	3	***4***	1	0	0	18
2	2	2	3	0	1	0	15
3	1	2	2	0	0	1	10
Objective	−48	−70	−96	0	0	0	0

Row	y_1	y_2	y_3	s_1	s_2	s_3	RHS
1	3/4	3/4	1	1/4	0	0	9/2
2	2	2	3	0	1	0	15
3	1	2	2	0	0	1	10
Objective	−48	−70	−96	0	0	0	0

Row	y_1	y_2	y_3	s_1	s_2	s_3	RHS
1	3/4	3/4	1	1/4	0	0	9/2
2	−1/4	−1/4	0	−3/4	1	0	3/2
3	−1/2	1/2	0	−1/2	0	1	1
Objective	24	2	0	24	0	0	432

⟶

The optimal value is 432 at $y_1 = 0$, $y_2 = 0$, and $y_3 = 9/2$.
The solution for the primal is in the objective row under the s_i.
The solution to the original (primal) is a minimum of 432 at
$x_1^* = 24$, $x_2^* = 0$, and $x_3^* = 0$.

CHAPTER 6

LINEAR PROGRAMMING – APPLICATION MODELS

EXERCISES CHAPTER 6

1. a) Let x_{ij} be the amount to ship from warehouse i to store j, $i = 1, 2, 3, 4$, and $j = 1, 2, \ldots, 6$

$$\begin{aligned}\text{Minimize}: Z = {} & 9x_{11} + 7x_{12} + 9x_{13} + 9x_{14} + 6x_{15} + 10x_{16} \\ & + 12x_{21} + 6x_{22} + 9x_{23} + 7x_{24} + 11x_{25} + 5x_{26} \\ & + 3x_{31} + 6x_{32} + 7x_{33} + 5x_{34} + 10x_{35} + 11x_{36} \\ & + 5x_{41} + 8x_{42} + 2x_{43} + 2x_{44} + 11x_{45} + 3x_{46}\end{aligned}$$

Subject to:

$$\begin{aligned}& x_{11} + x_{12} + x_{13} + x_{14} + x_{15} + x_{16} \le 9 \\ & x_{21} + x_{22} + x_{23} + x_{24} + x_{25} + x_{26} \le 2 \qquad \text{from} \\ & x_{31} + x_{32} + x_{33} + x_{34} + x_{35} + x_{36} \le 6 \qquad \text{warehouses} \\ & x_{41} + x_{42} + x_{43} + x_{44} + x_{45} + x_{46} \le 5\end{aligned}$$

Solutions Manual to Accompany Finite Mathematics: Models and Applications, First Edition.
Carla C. Morris and Robert M. Stark.

$$x_{11} + x_{21} + x_{31} + x_{41} \geq 6$$
$$x_{12} + x_{22} + x_{32} + x_{42} \geq 4$$
$$x_{13} + x_{23} + x_{33} + x_{43} \geq 4 \qquad \text{to}$$
$$x_{14} + x_{24} + x_{34} + x_{44} \geq 2 \qquad \text{stores}$$
$$x_{15} + x_{25} + x_{35} + x_{45} \geq 4$$
$$x_{16} + x_{26} + x_{36} + x_{46} \geq 2$$
$$x_{ij} \geq 0 \qquad \text{all } i, j$$

The objective value is 99 when $x_{12}^* = 4$, $x_{14}^* = 1$, $x_{15}^* = 4$, $x_{26}^* = 2$, $x_{31}^* = 6$, $x_{43}^* = 4$, $x_{44}^* = 1$ and all other x_{ij}^* are zero.

b) Replace $x_{11} + x_{21} + x_{31} + x_{41} \geq 6$ with $x_{11} + x_{21} + x_{31} + x_{41} \geq 4$. The objective value is 88 when $x_{12}^* = 3$, $x_{15}^* = 4$, $x_{26}^* = 2$, $x_{31}^* = 4$, $x_{32}^* = 1$, $x_{34}^* = 1$, $x_{43}^* = 4$, $x_{44}^* = 1$ and all other x_{ij}^* are zero.

3. Let X_{ij} be the amount of whiskey i in blend j
i = A, B, C corresponds to A, B, C
j = 1, 2, 3 corresponds to BH, OC, MC(Blue Hen, Old College, Modeler's Choice)

$X_{\text{A BH}} + X_{\text{A OC}} + X_{\text{A MC}}$ Amount of BH
$X_{\text{B BH}} + X_{\text{B OC}} + X_{\text{B MC}}$ Amount of OC
$X_{\text{C BH}} + X_{\text{C OC}} + X_{\text{C MC}}$ Amount of MC

So, the cost of the basic wines is

$$8.50(X_{\text{A BH}} + X_{\text{B BH}} + X_{\text{C BH}}) + 7.00(X_{\text{A OC}} + X_{\text{B OC}} + X_{\text{C OC}})$$
$$+5.50(X_{\text{A MC}} + X_{\text{B MC}} + X_{\text{C MC}})$$

Cost of whiskeys:

$$7(X_{\text{A BH}} + X_{\text{A OC}} + X_{\text{A MC}}) + 6(X_{\text{B BH}} + X_{\text{B OC}} + X_{\text{B MC}})$$
$$+5(X_{\text{C BH}} + X_{\text{C OC}} + X_{\text{C MC}})$$

The profit, is

$$Z = \text{Revenue} - \text{Cost of Whiskeys}$$

⟶

Subject to:

$$X_{\text{A BH}} + X_{\text{A OC}} + X_{\text{A MC}} \leq 2100$$
$$X_{\text{B BH}} + X_{\text{B OC}} + X_{\text{B MC}} \leq 800 \qquad \text{available whiskey}$$
$$X_{\text{C BH}} + X_{\text{C OC}} + X_{\text{C MC}} \leq 1700$$

Blue Hen blend

$$\frac{X_{\text{A BH}}}{X_{\text{A BH}} + X_{\text{B BH}} + X_{\text{C BH}}} \geq 0.40; \quad \frac{X_{\text{C BH}}}{X_{\text{A BH}} + X_{\text{B BH}} + X_{\text{C BH}}} \leq 0.30$$

Old College blend

$$\frac{X_{\text{A OC}}}{X_{\text{A OC}} + X_{\text{B OC}} + X_{\text{C OC}}} \geq 0.25; \quad \frac{X_{\text{C OC}}}{X_{\text{A OC}} + X_{\text{B OC}} + X_{\text{C OC}}} \leq 0.40$$

Master's Choice blend,

$$\frac{X_{\text{C MC}}}{X_{\text{A MC}} + X_{\text{B MC}} + X_{\text{C MC}}} \leq 0.60$$

$$\text{All}\, X_{ij} \geq 0 \quad i = \text{A, B, C} \quad j = \text{BH, OC, MC}$$

The maximum revenue is \$9500 when $X^*_{\text{A BH}} = 2100$, $X^*_{\text{B BH}} = 800$, and $X^*_{\text{C BH}} = 1242.857$ and all other $X^*_{ij} = 0$. Produce Blue Hen only.

5. The caterer has three options:

   1) Purchase x_j new napkins for use on j^{th} day at a \$/napkin.
   2) Use normal (full day) laundering of X_j soiled napkins at the end of the jth day at b \$/napkin and ready for use on day $j + 2$.
   3) Use rapid (overnight) service of Y_j napkins at the end of the jth day at c \$/napkin and ready for use on day $j + 1$.

   The linear program is

   $$\text{Minimize}: \; 5w_1 + 5w_2 + 5w_3 + 5w_4 + 5w_5 + 5w_6 + 5w_7 + x_1 + x_2 + x_3 + x_4 + x_5 + x_6 + x_7 + 1.5y_1 + 1.5y_2 + 1.5y_3 + 1.5y_4 + 1.5y_5 + 1.5y_6 + 1.5y_7$$

$\longrightarrow$

Subject to:

$$w_1 - k_1 = 250$$
$$w_2 + y_1 + k_1 - k_2 = 75$$
$$w_3 + y_2 + k_2 + x_1 - k_3 = 95$$
$$w_4 + y_3 + k_3 + x_2 - k_4 = 120$$
$$w_5 + y_4 + k_4 + x_3 - k_5 = 70$$
$$w_6 + y_5 + k_5 + x_4 - k_6 = 300$$
$$w_7 + y_6 + k_6 + x_5 - k_7 = 500$$
$$x_1 + y_1 + k_1 - w_1 = 0$$
$$x_1 + x_2 + y_2 + k_2 - w_1 - w_2 = 0$$
$$x_2 + x_3 + y_3 + k_3 - w_1 - w_2 - w_3 = 0$$
$$x_3 + x_4 + y_4 + k_4 - w_1 - w_2 - w_3 - w_4 = 0$$
$$x_4 + x_5 + y_5 + k_5 - w_1 - w_2 - w_3 - w_4 - w_5 = 0$$
$$x_5 + x_6 + y_6 + k_6 - w_1 - w_2 - w_3 - w_4 - w_5 - w_6 = 0$$

The minimum cost is \$3560 when

$$w_1^* = 325,\ w_3^* = 45,\ w_7^* = 130,\ w_2^* = w_4^* = w_5^* = w_6^* = 0$$
$$x_1^* = 250,\ x_2^* = 75,\ x_3^* = 95,\ x_4^* = 120,\ x_5^* = 70,\ x_6^* = x_7^* = 0$$
$$y_1^* = y_2^* = y_3^* = y_4^* = y_5^* = y_7^* = 0,\ y_6^* = 300$$
$$k_1^* = 75,\ k_3^* = 200,\ k_4^* = 155,\ k_5^* = 180,\ k_2^* = k_6^* = k_7^* = 0$$

CHAPTER 7

SET AND PROBABILITY RELATIONSHIPS

EXERCISES 7.1

1. a) $7 \notin \mathbf{A}$ True

b) $\{2, 3\} \subset \mathbf{A}$ True

c) $\{1\} \subset \mathbf{A}$ True

d) $\{2, 3, 4\} \subseteq \mathbf{A}$ False

e) $4 \notin \mathbf{A}$ False

3. a) To determine $n(\mathbf{A} \cap \mathbf{B})$, find $\mathbf{A} \cap \mathbf{B}$ and count the number of elements in this set. $\mathbf{A} \cap \mathbf{B} = \{2, 3, 5\}$ and, therefore,

$$n(\mathbf{A} \cap \mathbf{B}) = 3.$$

b) $(\mathbf{A} \cup \mathbf{B}) = \{1, 2, 3, 4, 5, 7\}$, so $n(\mathbf{A} \cap \mathbf{B}) = 6$.

c) $\mathbf{A}^c$ includes all elements not in $\mathbf{A}$ but are in the universal set $\mathbf{U}$.

$$\mathbf{A}^c = \{1, 4, 6, 8, 9\}, \text{so } n(\mathbf{A}^c) = 5.$$

⟶

Solutions Manual to Accompany Finite Mathematics: Models and Applications, First Edition.
Carla C. Morris and Robert M. Stark.

d) $\mathbf{A}^c \cup \mathbf{B} = \{1, 2, 3, 4, 5, 6, 8, 9\}$ and $(\mathbf{A}^c \cup \mathbf{B})^C = \{7\}$. Therefore,

$$n(\mathbf{A}^c \cup \mathbf{B})^c = 1.$$

e) $\mathbf{A} \cap \mathbf{B}^c = \{2, 3, 5, 7\} \cap \{6, 7, 8, 9\} = \{7\}$, so $n(\mathbf{A} \cap \mathbf{B}^c) = 1$.

5. a) $\mathbf{A} \cap \mathbf{B} = \{3, 5, 7\}$
b) $\mathbf{A} \cup \mathbf{B} = \{1, 2, 3, 4, 5, 6, 7\}$
c) $\mathbf{A}^c = \{2, 4, 6, 8, 9, 10\}$
d) $\mathbf{A}^c \cup \mathbf{B}^c = \{2, 4, 6, 8, 9, 10\} \cup \{1, 8, 9, 10\}$
$= \{1, 2, 4, 6, 8, 9, 10\}$
e) $\mathbf{A} \cap \mathbf{B}^c = \{1, 3, 5, 7\} \cap \{1, 8, 9, 10\} = \{1\}$

7. a) $\mathbf{A} = \{$Jan, Mar, May, July, Aug, Oct, Dec$\}$, so $n(\mathbf{A}) = 7$
b) $\mathbf{C} = \{$Apr, June, Sep, Nov$\}$, so $n(\mathbf{C}) = 4$
c) $\mathbf{B} \cup \mathbf{C} = \{$Feb, Apr, June, Sep, Nov$\}$, so $n(\mathbf{B} \cup \mathbf{C}) = 5$

9. Answers vary depending on the sets chosen. One example is:
Let $\mathbf{A} = \{1, 2, 3, 4\}$, $\mathbf{B} = \{3, 4, 5, 6, 7\}$, and $\mathbf{C} = \{1, 3, 5, 7, 9\}$ then $(\mathbf{A} \cup \mathbf{C}) \cap (\mathbf{B} \cup \mathbf{C}) = \{1, 2, 3, 4, 5, 7, 9\} \cap \{1, 3, 4, 5, 6, 7, 9\}$ $= \{1, 3, 4, 5, 7, 9\}$ and $(\mathbf{A} \cap \mathbf{B}) \cup \mathbf{C} = \{3, 4\} \cup \{1, 3, 5, 7, 9\}\}$ $= \{1, 3, 4, 5, 7, 9\}$

11. a) $\mathbf{A} \cap \mathbf{B} = \{1, 4, 9, 16, 25, 36, 49\}$
b) $\mathbf{A} \cap \mathbf{C} = \{2, 3, 5, 7, 11, 13, 17, 23, 29, 31, 37, 41, 43, 47\}$

13. One knows:

$$n(x_1) = (4/3)n(x_2) \text{ and } n(x_3) = 3n(x_2)$$

$$n(x_1) + n(x_2) + n(x_3) = (4/3)n(x_2) + n(x_2) + 3n(x_2) = 80$$

$$(16/3)n(x_2) = 80$$

$$n(x_2) = 15$$

Therefore, $n(x_1) = 20$, $n(x_2) = 15$, and $n(x_3) = 45$.

15. $\mathbf{A} = \{$Tom Sawyer, Huckleberry Finn, A Connecticut Yankee in King Arthur's Court$\}$

$\mathbf{B} = \{$The Black Cat, Cask of Amontillado, The Fall of the House of Usher, Pit and the Pendulum, Murders in the Rue Morgue$\}$

$\mathbf{C} = \{$Moby Dick, Billy Budd$\}$

So $n(\mathbf{A}) = 3$, $n(\mathbf{B}) = 5$, and $n(\mathbf{C}) = 2$.

17. Let **A** = read Time and **B** = read US News &World Report
$n(\mathbf{U}) = 150$, $n(\mathbf{A}) = 50$, $n(\mathbf{B}) = 90$, and $n(\mathbf{A} \cap \mathbf{B}) = 30$.

a) $n(\mathbf{A}\ \text{only}) = n(\mathbf{A}) - n(\mathbf{A} \cap \mathbf{B}) = 50 - 30 = 20$

b) $n(\mathbf{A} \cup \mathbf{B})^c = 150 - n(\mathbf{A} \cup \mathbf{B}) = 150 - [n(\mathbf{A}) + n(\mathbf{B}) - n(\mathbf{A} \cap \mathbf{B})]$
$= 150 - [50 + 90 - 30] = 40$

19. {OPST, OPTS, OSPT, OSTP, OTPS, OTSP, POST, POTS, PSOT, PSTO, PTOS, PTSO, SOPT, SOTP, SPOT, SPTO, STOP, STOP, TOPS, TOSP, TPOS, TPSO, TSOP, TSPO}

There are six words {OPTS, POST, POTS, SPOT, STOP, TOPS}

21. {H1, H2, H3, H4, H5, H6, T1, T2, T3, T4, T5, T6}. There are 12 or $2 \times 6 = 12$.

23. $3 \times 5 = 15$

25. $n(\mathbf{A} \cup \mathbf{B} \cup \mathbf{C}) =$
$n(\mathbf{A}) + n(\mathbf{B}) + n(\mathbf{C}) - n(\mathbf{A} \cap \mathbf{B}) - n(\mathbf{A} \cap \mathbf{C}) - n(\mathbf{B} \cap \mathbf{C})$
$+ n(\mathbf{A} \cap \mathbf{B} \cap \mathbf{C})$
Let $\mathbf{A} = \{1, 2, 3, 4, 5\}$, $\mathbf{B} = \{3, 4, 5, 6, 7\}$, and $\mathbf{C} = \{5, 7, 8, 9\}$ then
$(\mathbf{A} \cup \mathbf{B} \cup \mathbf{C}) = \{1, 2, 3, 4, 5, 6, 7, 8, 9\}$
$\mathbf{A} \cap \mathbf{B} = \{3, 4, 5\}$, $\mathbf{B} \cap \mathbf{C} = \{5, 7\}$, $\mathbf{A} \cap \mathbf{C} = \{5\}$, and
$\mathbf{A} \cap \mathbf{B} \cap \mathbf{C} = \{5\}$
Using, $n(\mathbf{A} \cup \mathbf{B} \cup \mathbf{C}) = 5 + 5 + 4 - 3 - 2 - 1 + 1 = 9$
(which is correct).

EXERCISES 7.2

1. Consider Venn diagrams with regions indicated by Roman Numerals. Using the Venn diagram with two overlapping circles,

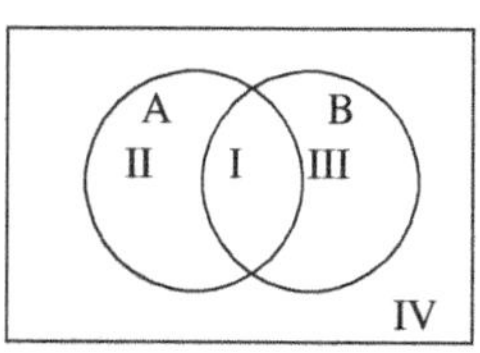

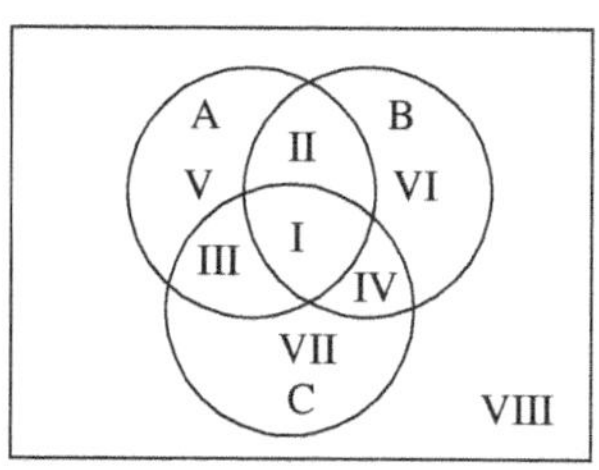

A ∪ **B** is represented by regions I, II, III and **B** ∪ **A** is regions I, II, III indicating that the commutative property for the union of two sets holds.

⟶

Likewise, **A** ∩ **B** is region I and so is **B** ∩ **A**, verifying the commutative property for intersections of sets.

Using the Venn diagram with three overlapping circles to establish the associative properties, one has

$$\mathbf{A} \cup (\mathbf{B} \cup \mathbf{C}) = \{I, II, III, V\} \cup \{I, II, III, IV, VI, VII\}$$
$$= \{I, II, III, IV, V, VI, VII\}$$
$$(\mathbf{A} \cup \mathbf{B}) \cup \mathbf{C} = \{I, II, III, IV, V, VI\} \cup \{I, III, IV, VII\}$$
$$= \{I, II, III, IV, V, VI, VII\}$$
$$\mathbf{A} \cap (\mathbf{B} \cap \mathbf{C}) = \{I, II, III, V] \cap \{I, IV\} = \{I\}$$
$$(\mathbf{A} \cap \mathbf{B}) \cap \mathbf{C} = \{I, II\} \cap \{I, III, IV, VII\} = \{I\}$$

To establish the distributive properties, one has

$$\mathbf{A} \cup (\mathbf{B} \cap \mathbf{C}) = (\mathbf{A} \cup \mathbf{B}) \cap (\mathbf{A} \cup \mathbf{C})$$
$$\{I, II, III, V\} \cup \{I, IV\}$$
$$= \{I, II, III, IV, V, VI\} \cap \{I, II, III, IV, V, VII\}$$
$$\{I, II, III, IV, V\} = \{I, II, III, IV, V\}$$

and

$$\mathbf{A} \cap (\mathbf{B} \cup \mathbf{C}) = (\mathbf{A} \cap \mathbf{B}) \cup (\mathbf{A} \cap \mathbf{C})$$
$$\{I, II, III, V\} \cap \{I, II, III, IV, VI, VII = \{I, II\} \cup \{I, III\}$$
$$\{I, II, III\} = \{I, II, III\}$$

3. a) B ⊂ A

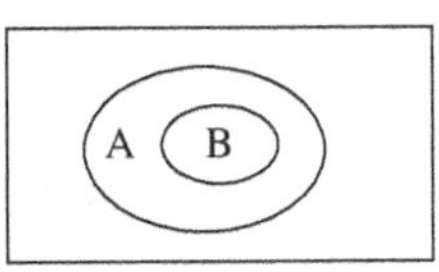

b) A ∩ B = ∅

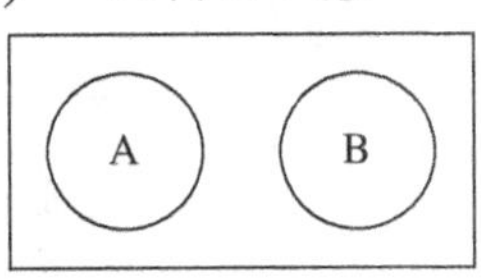

5. Using the Venn diagram with two overlapping circles
$(\mathbf{A} \cap \mathbf{B})^c = \{I\}^c = \{II, III, IV\}$ while
$\mathbf{A}^c \cap \mathbf{B}^c = \{III, IV\} \cap \{II, IV\} = \{IV\}$.

7. $(\mathbf{A}^c \cup \mathbf{B}^c) = \{III, IV\} \cup \{II, IV\} = \{II, III, IV\}$
So $n(\mathbf{A}^c \cup \mathbf{B}^c) = 20 + 18 + 7 = 45$

9. Let T = Time and US = US News & World Report. First, place 30 in the intersection. Use that 50 read Time and 90 read US News & World Report to yield 20 and 60, respectively. There are 150 individuals, so 40 are outside the sets.

⟶

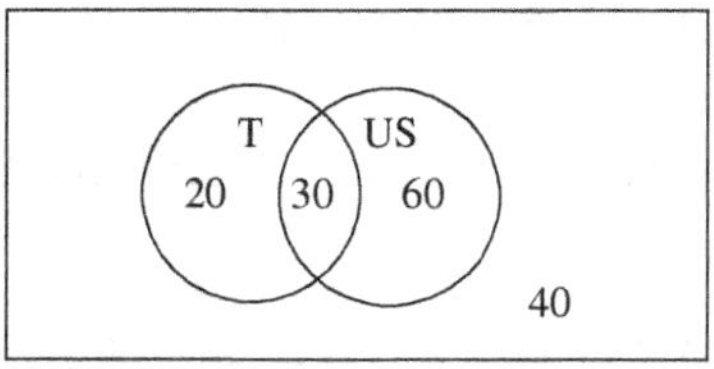

a) The Venn diagram shows 20 read Time only.

b) Forty read neither publication.

11. The Venn diagram is:

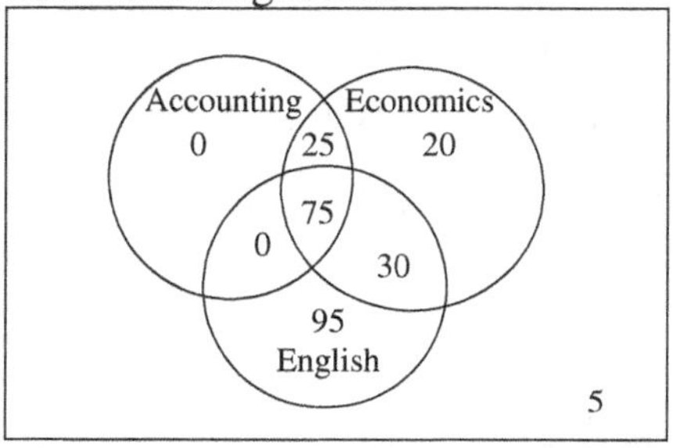

The diagram has 5 math students not taking any of the other three courses (region VIII).

13. The Venn diagram is:

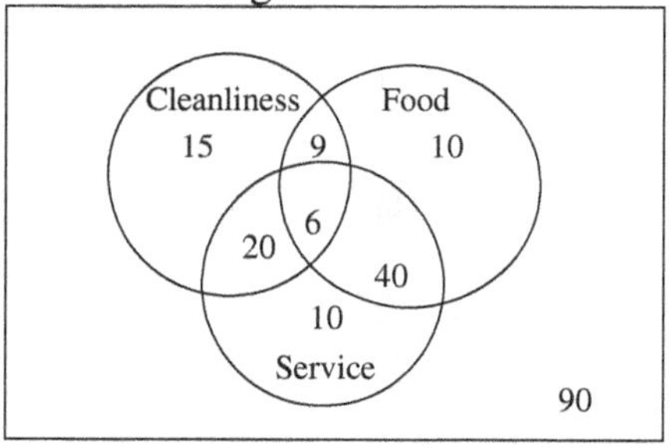

The diagram indicates that 35 (= 15 + 10 + 10) complained of a single issue (regions V, VI, and VII).

15. Three overlapping circles are used A antigen present, B antigen present, and Rh positive.

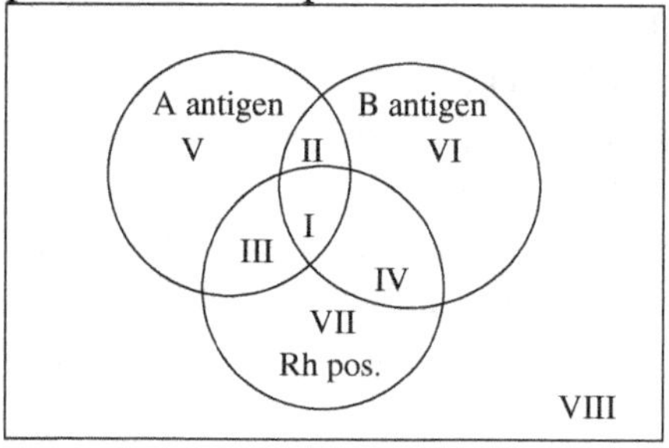

⟶

a) The diagram indicates the following blood types for each region

I AB^+	III A^+	V A^-	VII O^+
II AB^-	IV B^+	VI B^-	VIII O^-

b) The percentages of US populations with the four blood types are:

A 40%	AB 4%
B 11%	O 45%

c) There are extreme differences in blood types in the world. South America is almost entirely, type O, for example.

EXERCISES 7.3

1. There are $8 = (2^3)$ subsets including the empty set and the set itself

$$\{\ \}, \{X\}, \{Y\}, \{Z\}, \{X, Y\}, \{X, Z\}, \{Y, Z\}, \{X, Y, Z\}$$

3. There are 15 ways as shown subsequently (RR not possible)

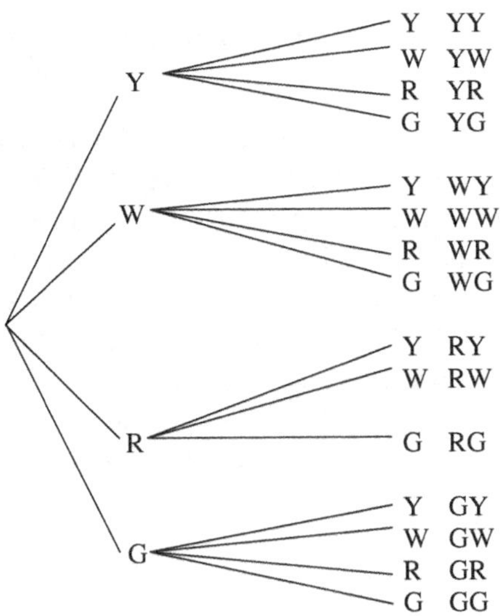

5. The tree diagram with the possibilities listed to the right is:

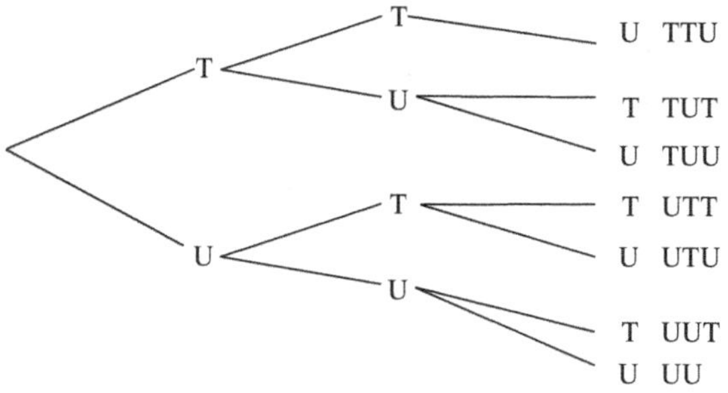

7. A "tree" depicts "improved reliability" followed by " improved durability". Since 32 of 50 have improved reliability, 18 do not. These numbers are on the first branch in the tree. There are 6 on the NN branch since they have neither improved reliability nor improved durability. Next, the YY branch has 15 (= 27 − 12)

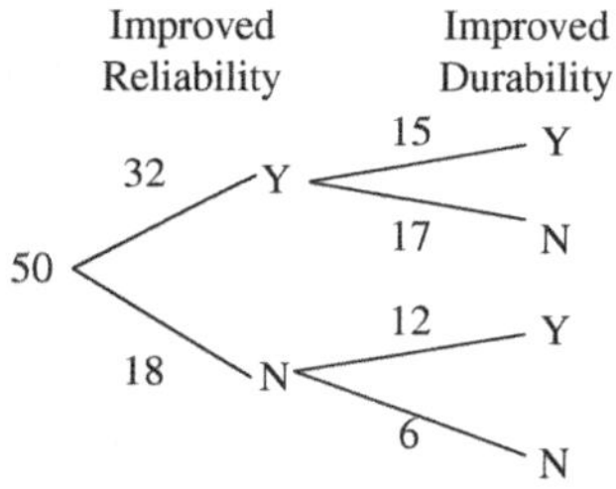

There are 15 with both improved reliability and improved durability.

9. First, consider three questions: 1) Did you drink iced tea? 2) Did you drink soda? and 3) Did you drink lemonade? Here, the YYY branch has 21 and the NNN branch has 0 to begin. Since 32 had iced tea and lemonade, the YYY and YNY branches must equal the total. Since YYY has 21 leaving YNY with 11. Using the same reasoning for the other pairs of drinks, the YYN branch has 12 and NYY branch 23. Since 18 had soda, only the NYN has this amount. The first split has 80 who drank iced tea, and, therefore, 70 did not. By working "back to front" from the lemonade question, branch numbers are determined.

⟶

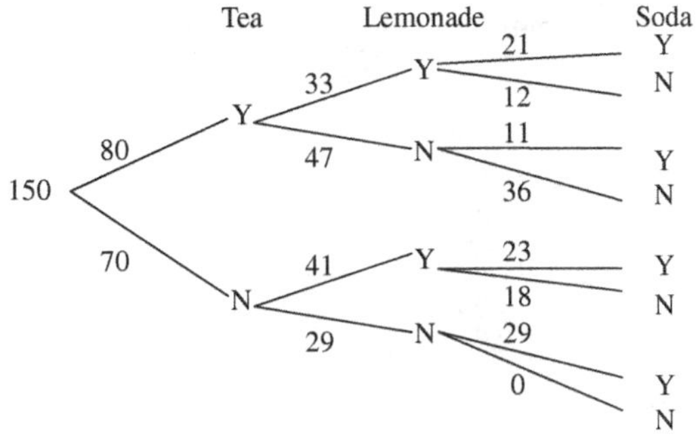

a) 29 had lemonade only b) 74 had soda.

11. One possible tree diagram is:

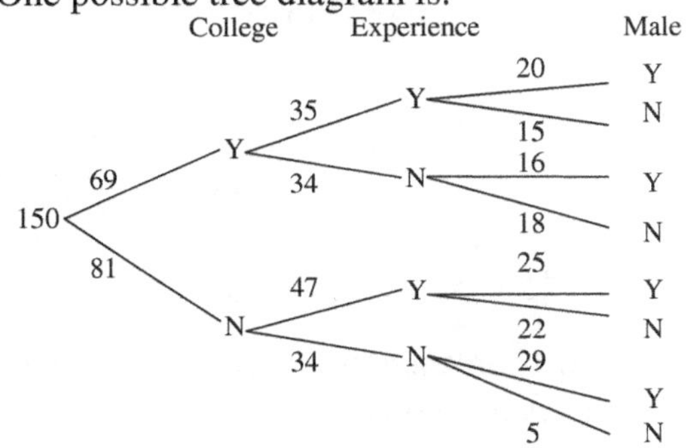

Female college graduates are the YNN and YYN branches. Therefore, there are $15 + 18 = 33$ female college graduates. The order of questions can be changed for other trees with the same result.

13. The tree and the 15 associated possibilities are:

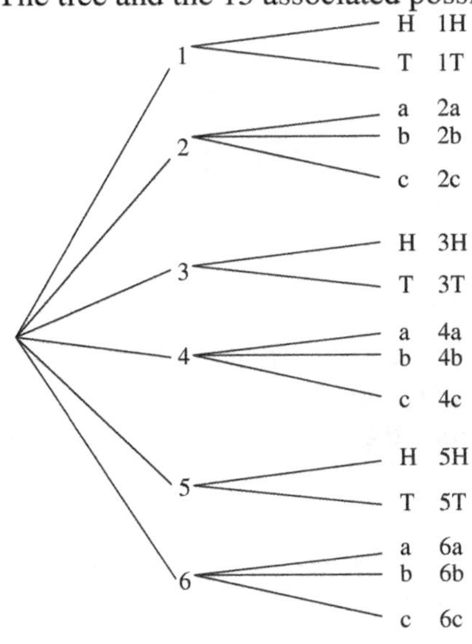

15. The tree and 16 possibilities are:

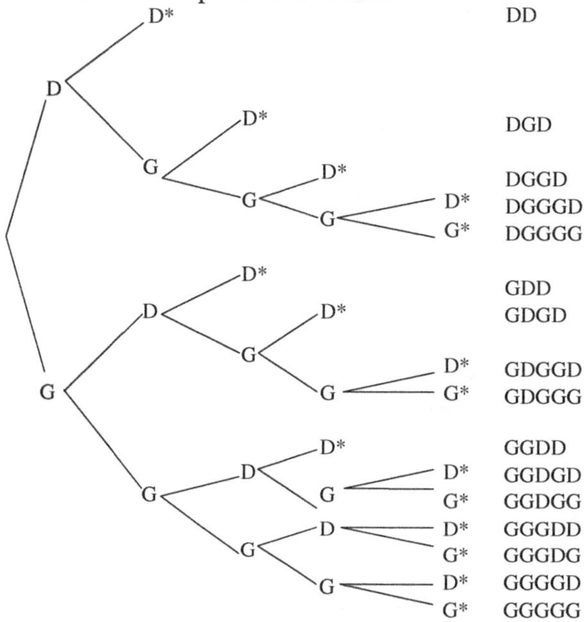

17. As a start, 57 are women, so 43 are men to total 100. The 23 male freshmen appear on the NY branch and remaining information is placed on the tree.

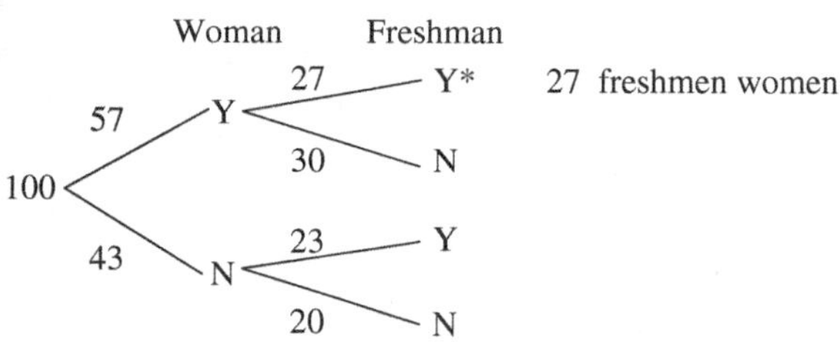

There are 27 freshmen women in the class of 100.

EXERCISES 7.4

1. There are five digits 2, 3, 6, 7, and 8 to form a three-digit odd number without repetition. The last digit determines if the number is odd or even. (Fill three slots in reverse: the third digit, second digit, and first digit) so

(two possibilities are 3 or 7) (any of four numbers) (any of three numbers).
Using the Multiplication Principle, there are $(2)(4)(3) = 24$ possibilities.

3. There are $(4)(2)(3) = 24$ possibilities.

5. a) $7! = 7 \cdot 6 \cdot 5 \cdot 4 \cdot 3 \cdot 2 \cdot 1 = 5040$
b) $8!/6! = 56$
c) $9!/0! = 362,880$
d) $10!/10! = 1$
e) $52!/(48!4!) = (52 \cdot 51 \cdot 50 \cdot 49)/4! = 270,725$

7. There are $\binom{52}{5} = 2,598,960$ five card poker hands possible.

9.
a) ${}_7C_2 = 21$
b) ${}_6C_5 = 6$
c) $\binom{9}{6} = 84$
d) $\binom{4}{1} = 4$

11. A permutation of seven colors taken five at a time yields ${}_7P_5 = 2520$ possible map colorings.

13. There are six numbers to form a three-digit number without repetition. One solution uses ${}_6P_3$ a permutation of six items taken three at a time, or 120. Fill three slots in the sequence (any of six numbers)(any of five remaining)(any of four remaining) or $(6)(5)(4) = 120$ possibilities.

15. There are $2^4 = 16$ outcomes. Among them are four possibilities. If there are three heads there must be only one tail. The tail could have been first, second, third, or fourth, indicating four possibilities.

17. $r!\ {}_nC_r = r!\ \dfrac{n!}{r!(n-r)!} = \dfrac{n!}{(n-r)!} = {}_nP_r$

19. There are two possibilities to consider. Either two of the three finalists are chosen and one wrong choice or all the three finalists are chosen yielding:

$$\binom{3}{2}\binom{7}{1} + \binom{3}{3}\binom{7}{0} = 3 \cdot 7 + 1 \cdot 1 = 21 + 1 = 22 \text{ ways.}$$

21. The 14 students are divided into two groups of four and a group of five, which results in $\dfrac{14!}{5!5!4!} = 252{,}252$ possible seating arrangements.

23. Fill six slots with letter letter digit digit digit digit.
Therefore, there are $(26)(26)(10)(10)(10)(10) = 6,760,000$ possible license plates in this case.

25. There is a 1/10,000 chance someone successfully enters your four digit PIN on the first attempt. The chance is about 99.9% that an alarm is set off by using a wrong PIN on four attempts.

$$(9999/10{,}000)(9998/10{,}000)(9997/10{,}000)(9996/10{,}000) \approx 0.999.$$

27. a) In an area code, there are $10^7 = 10{,}000{,}000$ possible seven-digit telephone numbers.

b) Using the rules for area codes cited, there are $(9)(2)(10) = 180$ possible area codes, and, therefore, $180(10^7) = 1{,}800{,}000{,}000$ possible telephone numbers.

29. The DJ has three options where four CDs are chosen that represent all three groups. They are

$$2\text{ R\&B, 1 Country, 1 Pop} = \binom{10}{2}\binom{15}{1}\binom{25}{1} = 16{,}875$$

$$1\text{ R\&B, 2 Country, 1 Pop} = \binom{10}{1}\binom{15}{2}\binom{25}{1} = 26{,}250$$

$$1\text{ R\&B, 1 Country, 2 Pop} = \binom{10}{1}\binom{15}{1}\binom{25}{2} = 45{,}000$$

Therefore, there are $16{,}875 + 26{,}250 + 45{,}000 = 88{,}125$ possibilities.

31. There are four items for a circular permutation, so $(4-1)! = 6$.

33. There are 75 pieces of mail to freely distribute to 50 homes yielding

$$\binom{50+75-1}{50} = \binom{124}{50} \approx 1.5 \times 10^{35}$$

EXERCISES 7.5

1. When a pair of dice is thrown, the sum is 2 through 12, so the sample space is {2, 3, 4, 5, 6, 7, 8, 9, 10, 11, 12}.

3. a) {face, ace, number card} is a valid sample space.

b) {1head, 2 heads} does not include the possibility of 0 heads, so it is not a valid sample space.

c) {number less than 6, even number} not mutually exclusive events, so it is not a valid sample space.

5. $\frac{3}{6} + \frac{3}{6} - \frac{1}{6} = \frac{5}{6}$

7. Since **A** and **B** are mutually exclusive, $P(\mathbf{A} \cup \mathbf{B}) = P(\mathbf{A}) + P(\mathbf{B})$.
One also knows that $P(\mathbf{B}) = 1 - P(\mathbf{B}^c) = 0.7$, so
$P(\mathbf{A} \cup \mathbf{B}) = 0.2 + 0.7 = 0.9$

9. See Chapter notes. Answers vary by class size.

11. There are six possibilities (1, 1, 1), (1, 3, 1), (3, 1, 1), (1, 2, 2), (2, 1, 2), and (2, 2, 1). The probability is $\frac{6}{216} = \frac{1}{36}$

13. The probability of winning the NJ lotto is $\frac{1}{\binom{49}{6}} = \frac{1}{13{,}983{,}816}$

15. a) $\frac{39}{39+27} = \frac{39}{66}$ b) 5/39 c) $2/66 = 1/33$

d) $P(\text{Gospel}/\text{New Testament})$

$$= \frac{P(\text{Gospel} \cap \text{New Testament})}{P(\text{New Testament})} = \frac{4/66}{27/66} = \frac{4}{27}$$

17. a) 24/65 b) 0/65 c) 6/65 (GA and VA are among the first ten)

19. a) 7% are Rh-negative

b) $P(\text{O}/\text{Rh} - \text{Pos})$

$$= \frac{P(\text{O} \cap \text{Rh} - \text{Pos})}{P(\text{Rh} - \text{Pos})} = \frac{0.38}{0.38 + 0.34 + 0.09 + 0.03}$$

$$= \frac{38}{84} \approx 0.4524$$

c) The Universal Donor is O^-, so 7%

d) The Universal Receiver is AB^+, so 3%

21. $P(\text{sum } 7/\text{sum} \le 8) = \frac{P(\text{sum } 7 \cap \text{sum} \le 8)}{P(\text{sum} \le 8)} = \frac{6/36}{26/36} = \frac{6}{26} = \frac{3}{13}$

23. $P(\text{credit}/\text{not MC}) = \frac{P(\text{credit} \cap \text{not MC})}{P(\text{not MC})} = \frac{0.3}{0.7} = \frac{3}{7}.$

25. a) 32%

b) 25% + 1% = 26%

c) $P(\text{opposes}/\text{doesn't support}) = \frac{P(\text{opposes} \cap \text{doesn't support})}{P(\text{doesn't support})}$

$$= 0.39/0.43 \approx 0.907$$

27. There are five democrats and two Republicans to form a subcommittee of two. For "at least one democrat", either one or two Democrats are chosen.

⟶

$P(\text{both Dem}/\text{at least 1 Dem})$

$$= \frac{P(\text{Both } D \cap \text{at least 1 Dem})}{P(\text{at least 1 Dem})} = \frac{\binom{5}{2}\binom{2}{0}}{\binom{5}{1}\binom{2}{1} + \binom{5}{2}\binom{2}{0}} = \frac{10}{10+10} = \frac{1}{2}$$

29. There is some controversy over the Shakespearean sonnets and how many are actually to the "dark lady." Assuming that sonnets 127–154 are to the "dark lady", there are 28 sonnets to consider.

a) P(dark lady) = 28/154

b) $P(\text{dark lady}/\text{not } 1^{\text{st}}\ 100) = \dfrac{P(\text{dark lady} \cap \text{not } 1^{\text{st}}\ 100)}{P(\text{not } 1^{\text{st}}\ 100)}$

$$= \frac{28/154}{54/154} = \frac{28}{54} \approx 0.519$$

c) $P(\text{dark lady}/\text{perfect square}) = \dfrac{P(\text{dark lady} \cap \text{perfect square})}{P(\text{perfect square})}$

$$= \frac{1/154}{12/154} = \frac{1}{12} \approx .083.$$

31. The probabilities are given out of 52 cards. P(draw an ace) = 4/52 and P(draw a black card) = 26/52.

$$P(\text{ace} \cap \text{black}) = \frac{P(\text{ace} \cap \text{black})}{P(\text{black})} = \frac{2/52}{26/52} = \frac{2}{26} = \frac{1}{13}$$

Since $P(\text{ace}) = 4/52 = 1/13$, the two events are independent.

33. a) There were 36 Nascar races in 2003 and Ryan Newman won 8 of these, so the probability is 8/36.

b) 3/36 c) 5/36

35. a) All capital letters are needed to have the darkest skin shade each of which has a 1/2 probability, so $(1/2)^6 = 1/64$.

b) All lower case letters are needed, and using the same logic as in part a), the probability is again 1/64.

c) One capital requires one of three cases having a capital A, capital B, or capital C with the remaining letters lower case.

$$\underbrace{(1/2)(1/4)(1/4)}_{\text{Capital A}} + \underbrace{(1/4)(1/2)(1/4)}_{\text{Capital B}} + \underbrace{(1/4)(1/4)(1/2)}_{\text{Capital C}} = 3/32$$

EXERCISES 7.6

1. A tree lists possibilities. Let R denote a "red card is drawn" and B a "black card is drawn." The stage-wise probabilities appear on the branches.

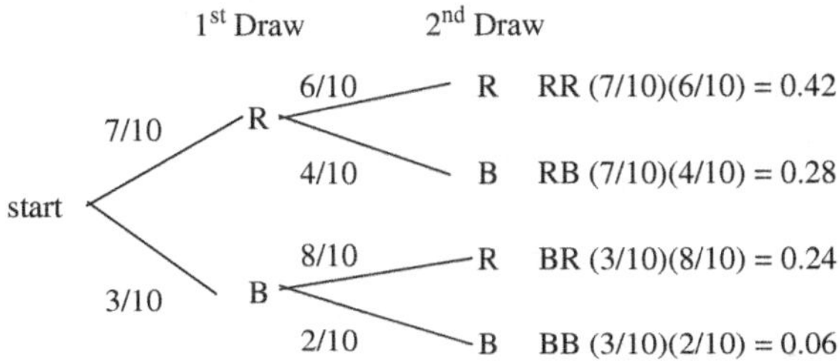

a) The second card is red if RR or BR and these branches yield $0.42 + 0.24 = 0.66$.

b) Both cards are black given they are same color and have probability

$$P(\text{Black}/\text{same}) = \frac{P(\text{black\&same})}{P(\text{same})} = \frac{P(\text{black})}{P(\text{same})} = \frac{0.06}{0.48} = \frac{1}{8}$$

Note:
$P(same) = P(both\ red) + P(both\ black) = 0.42 + 0.06 = 0.48$.

3. The tree diagram depicts the possibilities.

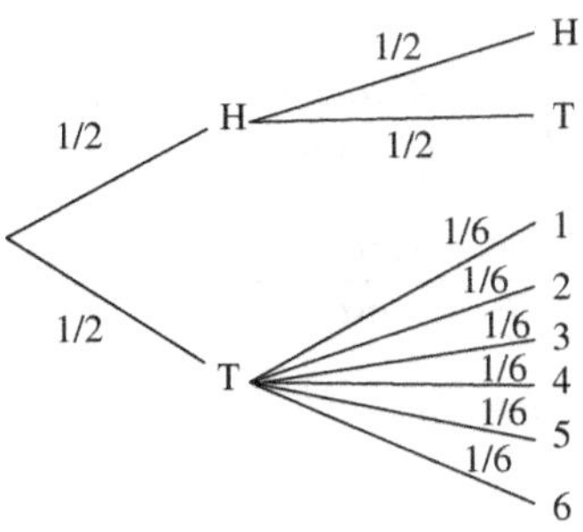

a) $P(\text{exactly one head}) = P(\text{HT}) = (1/2)(1/2) = 1/4$.

b) $P(\text{two or more}) = (1/2)(5/6) = 5/12$.

5. The tree diagram is an aid.

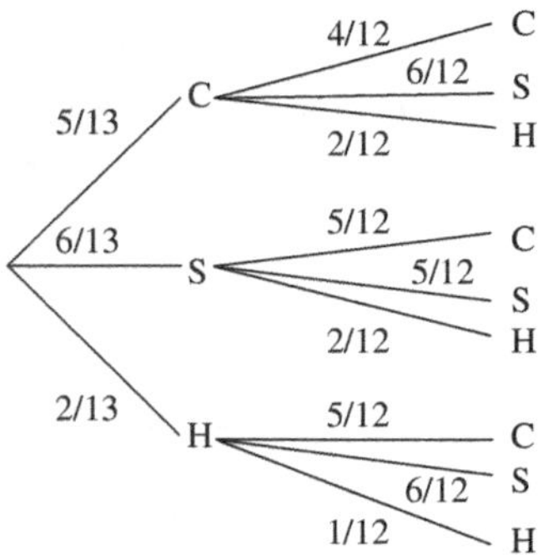

a) $P(\text{both spade}) = (6/13)(5/12) = 5/26$

b) $P(\text{both spade}/1^{\text{st}}\text{ spade}) = \dfrac{P(\text{SS} \cap 1^{\text{st}}\text{ spade})}{P(1^{\text{st}}\text{ spade})} = \dfrac{5/26}{6/13} = 5/12$

c) $P(\text{one } C/1^{\text{st}}\ \text{H}) = \dfrac{P(\text{one } C \cap 1^{\text{st}}\text{H})}{P(1^{\text{st}}\ \text{H})} = \dfrac{(2/13)(5/12)}{(2/13)} = 5/12$

7. If a couple has three children, there are 2^3 or 8 possibilities. "Having at least two girls" means that either two of the three children are girls or that all three are girls. The probability is

$$\frac{\binom{3}{2}+\binom{3}{3}}{2^3} = \frac{3+1}{8} = \frac{4}{8} = \frac{1}{2}.$$

9. "Not correctly filled" means the cans are either over- or underfilled. The probability of this occurring is 25%. Therefore,

$$P(\text{underfilled}/\text{not correctly filled}) = \frac{P(\text{under} \cap \text{not correct})}{P(\text{not correct})} = \frac{0.13}{0.25} = 0.52$$

11. Let A = ace, F = face, and N = number card.

a) To determine P(at least 1 F/ no A) requires finding P(no A). This can happen in four ways FF, FN, NF, and NN. The probability is $(12/52)(11/51) + (12/52)(36/51) + (36/52)(12/51) + (36/52)(35/51) = 2256/2652$
(While this can be simplified, the format is easier to track probabilities.) Next, determine the specific case(s) where there is "at least one F given that there are no A." There are three cases FF, FN, and NF $= (132 + 432 + 432)/2652 = 996/2652$.

⟶

$$P(\text{at least } 1\ \ F/\text{no } A) = \frac{P(\text{at least } 1\ \ F \cap \text{no } A)}{P(\text{no } A)}$$
$$= \frac{996/2652}{2256/2652} = \frac{996}{2256} = \frac{83}{188}$$

b) To determine P(both N / at least 1 N), use a similar reasoning. The events with "at least 1 N" are AN, FN, NA, NF, and NN. The three cases NA, NF, and NN can be thought of as a single case where an N is chosen first with probability 36/52. The other two cases have probability (4/52)(36/51) and (12/52)(36/51). Therefore, $P(\text{at least } 1\ N) = 2412/2652$. The specific case for the numerator is NN with probability 1260/2652 yielding $P(\text{both } N/\text{at least } 1\ N) = 1260/2412 = 35/67$.

13. a) $P(\mathbf{A}) = 0.6$, $P(\mathbf{B}) = 0.9$, $P(\mathbf{A}\,\text{pin}) = 0.8$, $P(\mathbf{B}\,\text{pin}) = 0.5$, as tests are independent

$$P(\mathbf{A} \cap \mathbf{B}) = (0.6)(0.9) = 0.54$$

$$P(\mathbf{A} \cup \mathbf{B}) = P(\mathbf{A}) + P(\mathbf{B}) - P(\mathbf{A} \cap \mathbf{B}) = 0.6 + 0.9 - 0.54 = 0.96$$

A Venn diagram can help.

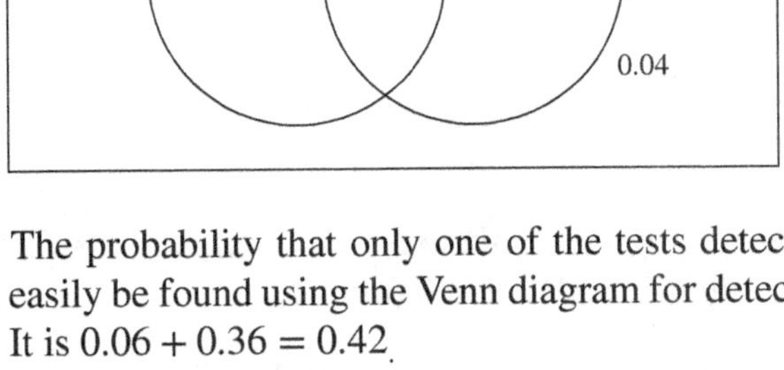

b) The probability that only one of the tests detects the disease can easily be found using the Venn diagram for detection probabilities. It is $0.06 + 0.36 = 0.42$.

c) To accurately pinpoint location, there are three cases. Only test A detects disease (D) and accurately pinpoints location (L). Only test B detects disease and accurately pinpoints location. Finally, both

⟶

tests detect the disease and, therefore, certain to pinpoint the location.

$$(0.06)(0.8) + (0.36)(0.5) + (0.54)(1) = 0.768$$

d) The probability that only test B detects and pinpoints a disease given a person identified as having a disease and its location is pinpointed, is determined using the information from part c)

$$\frac{P(\mathbf{B})P(\mathbf{D\&L}/\mathbf{B})}{P(\mathbf{A})P(\mathbf{D\&L}/\mathbf{A}) + P(\mathbf{B})P(\mathbf{D\&L}/\mathbf{B}) + P(\mathbf{A\&B})P(\mathbf{D\&L}/\mathbf{A\&B})}$$
$$\frac{(0.36)(0.5)}{0.768} = \frac{0.18}{0.768} = 0.234375$$

15. Using Bayes' Rule,

$$\begin{aligned} P(\mathbf{C}/\text{def}) &= \frac{P(\text{def}/\mathbf{C})P(\mathbf{C})}{P(\text{def}/\mathbf{A})P(\mathbf{A}) + P(\text{def}/\mathbf{B})P(\mathbf{B}) + P(\text{def}/\mathbf{C})P(\mathbf{C})} \\ &= \frac{(0.05)(0.45)}{(0.02)(0.30) + (0.03)(0.25) + (0.05)(0.45)} \\ &= \frac{0.0225}{0.0360} = \frac{5}{8} \end{aligned}$$

17. The question of flaws is a good starting point for the tree diagram. Experience shows that 12% of items are flawed and there is a 98% probability that the machine will detect this flaw but also an 8% chance that it will falsely indicate a flaw.

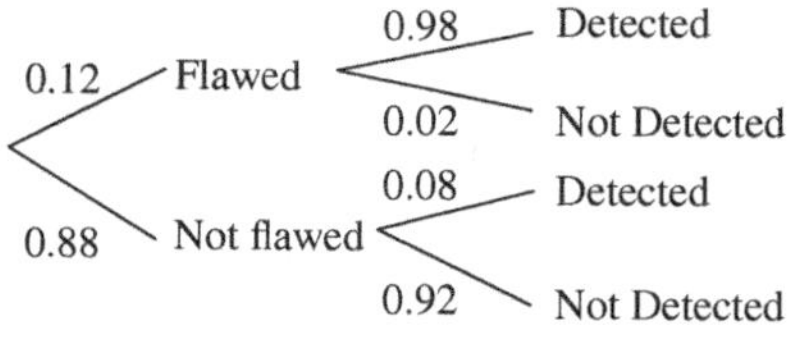

Therefore, the probability that an item is actually not flawed when the instrument indicates that it is flawed is:

$$P(\text{not flawed}/\text{flaw detected}) = \frac{(0.88)(0.08)}{(0.12)(0.98) + (0.88)(0.08)} \approx 0.374$$

19. This tree can be used or apply Bayes' Rule.

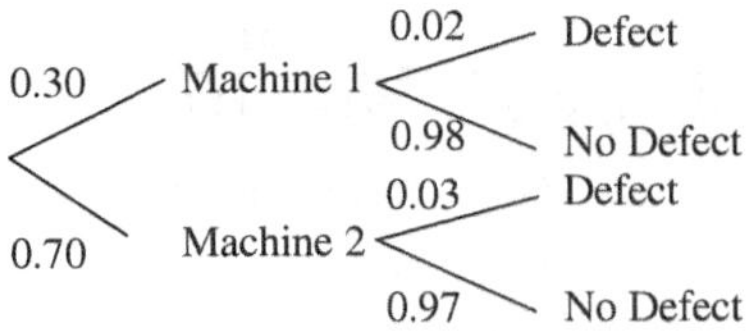

$$P(\text{M1}/\text{def}) = \frac{P(\text{def}/\text{M1})P(\text{M1})}{P(\text{def}/\text{M1})P(\text{M1}) + P(\text{def}/\text{M2})P(\text{M2})}$$

$$= \frac{(0.02)(0.3)}{(0.02)(0.3) + (0.03)(0.7)} = 0.\overline{2}$$

21. A tree diagram for this situation is:

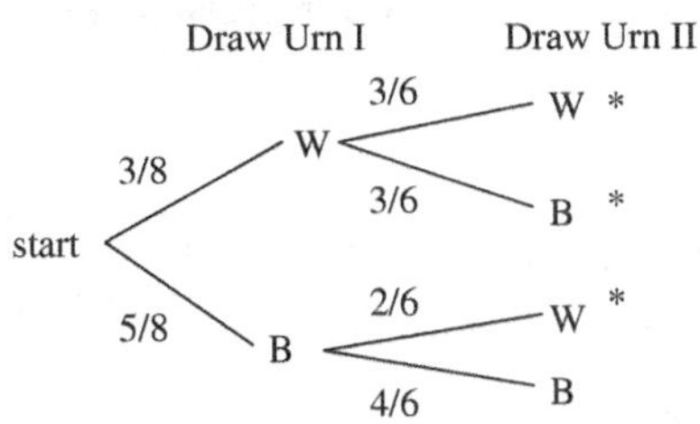

a) The probability that the ball drawn from Urn II is black is found on the WB and BB branches. So $(3/8)(3/6) + (5/8)(4/6) = 29/48$.

b) The probability that two white marbles are drawn given at least one is white is $P(2\text{W}/\text{at least 1W}) = \dfrac{P(2\text{W} \cap \text{at least 1W})}{P(\text{at least 1W})}$. There are three cases where at least 1W occurs denoted with *.

$$\frac{(3/8)(3/6)}{(3/8)(3/6) + (3/8)(3/6) + (5/8)(2/6)} = \frac{9/48}{28/48} = \frac{9}{28}$$

23. This tree can be used or apply Bayes' Rule.

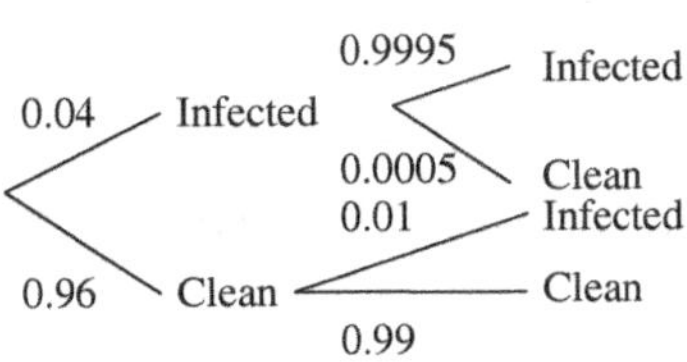

The probability that an e-mail found to be clean is actually infected is

$$\frac{(0.04)(0.0005)}{(0.04)(0.0005)+(0.96)(0.99)}=\frac{0.00002}{0.95042}$$
$$=0.000021$$

25. This tree can be used or apply Bayes' Rule.

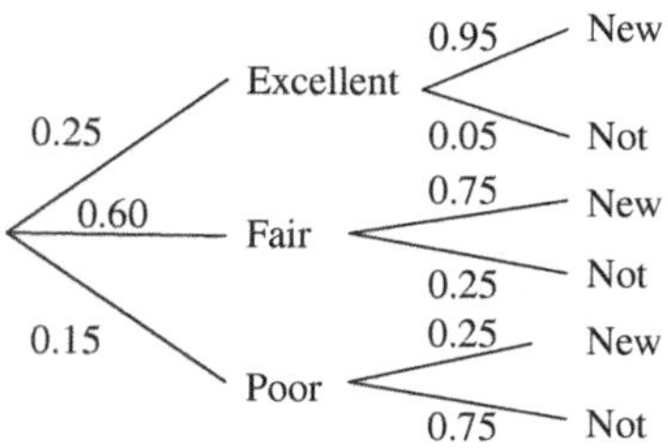

The probability that the credit rating was not poor given a new car was purchased is:

$P(\text{not poor}/\text{new}) =$

$$\frac{P(\text{new}/\text{excellent})P(\text{excellent})+P(\text{new}/\text{fair})P(\text{fair})}{P(\text{new}/\text{excellent})P(\text{excellent})+P(\text{new}/\text{fair})P(\text{fair})+P(\text{new}/\text{poor})P(\text{poor})}$$

$$=\frac{(0.95)(0.25)+(0.75)(0.60)}{(0.95)(0.25)+(0.75)(0.60)+(0.25)(0.15)}$$

$$=\frac{0.6875}{0.7250}\approx 0.9483$$

27. The tree can be used or apply Bayes' Rule.

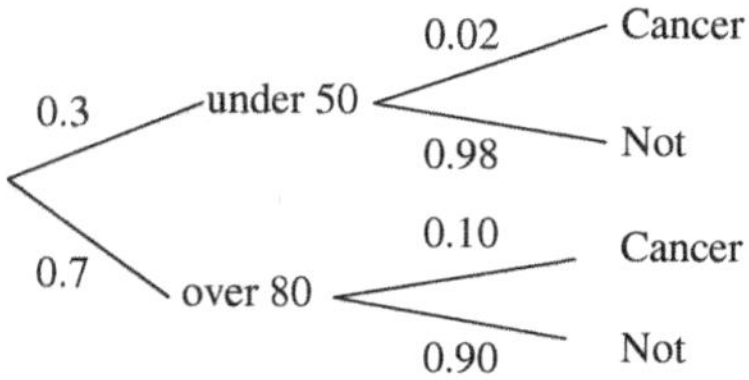

The probability that a woman is younger than 50 given she has developed breast cancer is:

$$\frac{(0.30)(0.02)}{(0.30)(0.02)+(0.7)(0.10)}=\frac{0.006}{0.076}\approx 0.07895.$$

SUPPLEMENTARY EXERCISES CHAPTER 7

1. a) True b) False $\mathbf{B} \cup \mathbf{C} = [1, 2, 3, 4, 5\}$ c)True

3. The sample space is:

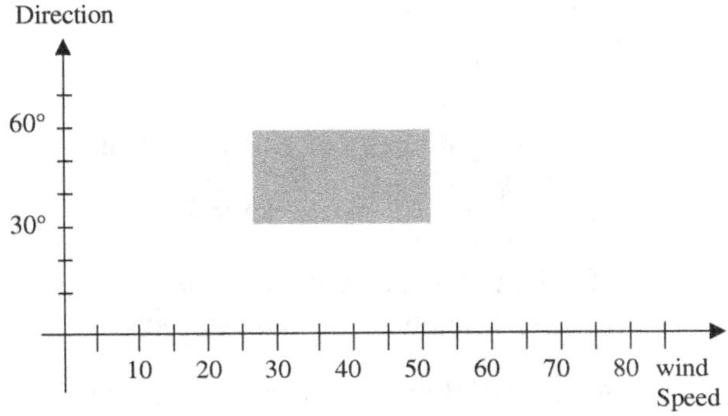

5. $\mathbf{A} \cap \mathbf{B} = \{100, 400\}$

7. The tree diagram is:

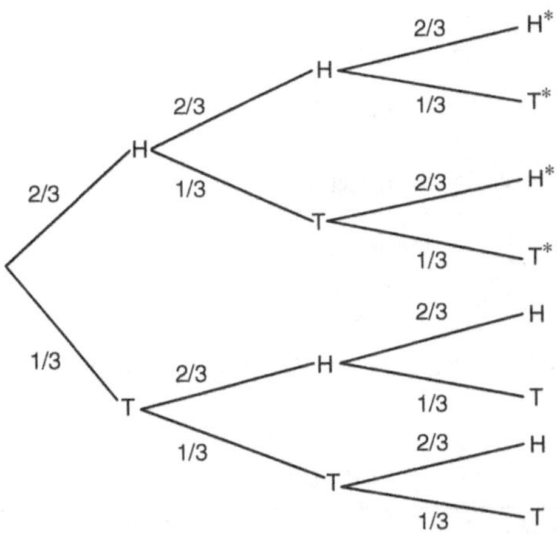

There are three cases where there are two heads and one with three heads. The probability is $(3)\left(\frac{2}{3}\right)^2\left(\frac{1}{3}\right)+\left(\frac{2}{3}\right)^3=\frac{20}{27}$

9. Imagine filling three positions. The gold medal can be won by any of the eight Olympians. The silver medal by any of the remaining seven, and the bronze by any of the remaining six to give (8)(7)(6) = 336 ways. Also three of eight Olympians can be chosen with order, a permutation, so ${}_8P_3 = 336$.

11. The six people can be thought of as a "group of one," "a group of two," and "a group of three" like objects according to the size of bedroom they are assigned. Therefore, there are $\dfrac{6!}{1!2!3!} = 60$ ways.

13. There are two cases to consider. One, "one of the six nickels and two of the three dimes are chosen." The other, "two of the six nickels and one of the three dimes are chosen." The result is

$$\binom{6}{1}\binom{3}{2} + \binom{6}{2}\binom{3}{1} = 6(3) + 15(3) = 63$$

15. There are six subsets with one element and fifteen that contain two elements as given by: $\binom{6}{1} + \binom{6}{2} = 6 + 15 = 21$

17. The probability that a box will not contain a defective fuse can be determined from the three branches without asterisks.

$P(\mathbf{A})P(\text{No defect}/\mathbf{A}) + P(\mathbf{B})P(\text{no defect}/\mathbf{B}) + P(\mathbf{C})\ \ P(\text{no defect}/\mathbf{C})$
$= (0.30)(0.98) + (0.50)(0.95) + (0.20)(0.90) = 0.949$

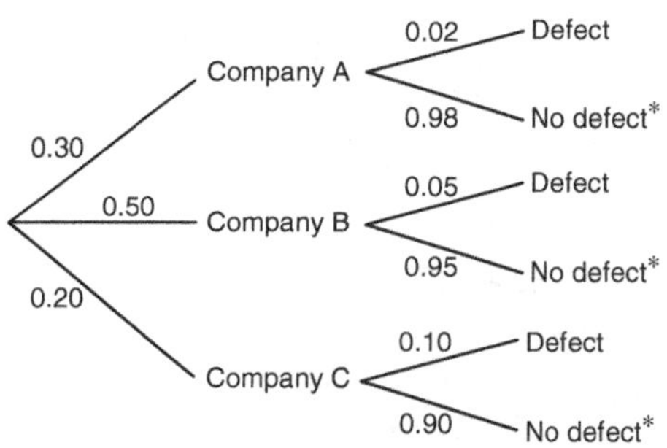

If a randomly chosen box has a defective, the probability that it came from Company B is:

$$\begin{aligned}P(\mathbf{B}/\text{def}) &= \frac{P(\text{def}/\mathbf{B})P(\mathbf{B})}{P(\text{def}/\mathbf{A})P(\mathbf{A}) + P(\text{def}/\mathbf{B})P(\mathbf{B}) + P(\text{def}/\mathbf{C})P(\mathbf{C})}\\ &= \frac{(0.05)(0.50)}{(0.02)(0.30) + (0.05)(0.50) + (0.10)(0.20)}\\ &= \frac{0.025}{0.051} \approx 0.49\end{aligned}$$

CHAPTER 8

RANDOM VARIABLES AND PROBABILITY DISTRIBUTIONS

EXERCISES 8.1

1. a) continuous b) discrete c) discrete d) continuous e) continuous

3. Economists are interested in the employment percentage in a city, the unemployment percentage in the United States last month, or the percentage change in the daily stock market. (Answers vary)

5. The probabilities must sum to unity

a) $1.0 - 0.2 - 0.1 - 0.4 = 0.3$

b) $1.0 - 0.05 - 0.35 - 0.25 - 0.20 = 0.15$

c) $1.0 - (1/16) - (1/8) - (3/16) - (5/16) - (1/8) = 3/16$

7. The six faces of a fair die have equal probability

x	1	2	3	4	5	6
$P(x)$	1/6	1/6	1/6	1/6	1/6	1/6

Solutions Manual to Accompany Finite Mathematics: Models and Applications, First Edition.
Carla C. Morris and Robert M. Stark.

9. The expected value in Exercise 8.1.5a is:

$$E(x) = 10(0.20) + 20(0.10) + 30(0.30) + 40(0.40) = 29.0$$

The expected value in Exercise 8.1.5b is:

$$E(x) = 1(0.05) + 5(0.35) + 9(0.25) + 12(0.20) + 20(0.15) = 10.05$$

The expected value in Exercise 8.1.5c is:

$$E(x) = -3(1/16) + 5(1/8) + 7(3/16) + 8(5/16) + 10(3/16) + 25(1/8) = 9.25$$

11. The face sum is from 2 to 12, with probability distribution for x:

x	2	3	4	5	6	7	8	9	10	11	12
$P(x)$	1/36	2/36	3/36	4/36	5/36	6/36	5/36	4/36	3/36	2/36	1/36

The expected value is

$$2(1/36) + 3(2/36) + \cdots + 12(1/36) = 252/36 = 7$$

13. The distribution is:

x	100,000	60,000
$P(x)$	0.75	0.25

$E(x) = 100{,}000(0.75) + 60{,}000(0.25) = 90{,}000$ without buying insurance. If the insurance is purchased, you anticipate $100{,}000 - 15{,}000 = 85{,}000$. Do not purchase the insurance as it is only worth \$10,000 on average.

15. The distribution for winnings not including the initial dollar for the ticket is:

x	1,500,000	500	50	0
$P(x)$	$\frac{\binom{6}{6}\binom{33}{0}}{\binom{39}{6}} = \frac{1}{3{,}262{,}623}$	$\frac{\binom{6}{5}\binom{33}{1}}{\binom{39}{6}} = \frac{198}{3{,}262{,}623}$	$\frac{\binom{6}{4}\binom{33}{2}}{\binom{39}{6}} = \frac{7920}{3{,}262{,}623}$	$\frac{3{,}254{,}504}{3{,}262{,}623}$

⟶

The expected value is $\dfrac{1,995,000}{3,262,623} \approx 0.6115$. However a dollar was paid, so there is an expected loss of about 39 cents.

17. The probability when a six is twice as likely yields:

x	1	2	3	4	5	6
$P(x)$	1/7	1/7	1/7	1/7	1/7	2/7

$$E(x) = 1(1/7) + 2(1/7) + \cdots + 6(2/7) = 27/7 \approx 3.857$$

19. Washington, Lincoln, Hamilton, Jackson, Grant, and Franklin appear on the \$1, \$5, \$10, \$20, \$50, and \$100 bills, respectively.

x	1	5	10	20	50	100
$P(x)$	6/20	3/20	4/20	4/20	2/20	1/20

Therefore,

$$E(x) = 1(6/20) + 5(3/20) + \cdots + 100(1/20) = \frac{341}{20} = \$17.05$$

EXERCISES 8.2

1. a) $P(x = 2) = \binom{3}{2}(0.4)^2(0.6) = 3(0.16)(0.16) = 0.288$

b) $P(x = 3) = \binom{4}{3}\left(\frac{1}{3}\right)^3\left(\frac{2}{3}\right) = 4\left(\frac{1}{27}\right)\left(\frac{2}{3}\right) = \frac{8}{81} = 0.09877$

c) $P(x = 1) = \binom{5}{1}(0.3)^1(0.7)^4 = 5(0.3)(0.2401) = 0.36015$

3. a) $P(x = 10) = \binom{10}{10}(0.9)^{10}(0.1)^0 = (0.9)^{10} = 0.348678$

b) $P(x = 20) = \binom{20}{20}(0.9)^{20}(0.1)^0 = (0.9)^{20} = 0.1215767$

5. The probability of rolling a "5" in a die toss is 1/6 and $n = 4$ here. There are two cases, $x = 2$ or $x = 3$. Therefore,

$$P(x = 2) + P(x = 3) = \binom{4}{2}\left(\frac{1}{6}\right)^2\left(\frac{5}{6}\right)^2 + \binom{4}{3}\left(\frac{1}{6}\right)^3\left(\frac{5}{6}\right)^1$$

$$= \frac{150 + 20}{1296} = 0.1312$$

7. a) $P(x=0) = \binom{5}{0}\left(\frac{1}{4}\right)^0\left(\frac{3}{4}\right)^5 = 1(1)\left(\frac{243}{1024}\right) = \frac{243}{1024} = 0.2373$

b) "At most three correct" means that 0, 1, 2, or 3 are correct. Calculate the sum of these four probabilities or use the complementary event: "1 – probability of 4 or 5 correct." Then,

$$1 - P(x=4) - P(x=5) = 1 - \binom{5}{4}\left(\frac{1}{4}\right)^4\left(\frac{3}{4}\right)^1 - \binom{5}{5}\left(\frac{1}{4}\right)^5\left(\frac{3}{4}\right)^0$$

$$= 1 - \frac{15}{1024} - \frac{1}{1024} = \frac{1008}{1024} = 0.984375$$

c) $P(x=5) = \binom{5}{5}\left(\frac{1}{4}\right)^5\left(\frac{3}{4}\right)^0 = \frac{1}{1024} = 0.0009766$

9. To earn a passing grade, 6, 7, 8, 9, or 10 correct answers are necessary. The probability of guessing a correct answer on any question is 1/2. Using the binomial distribution with $n = 10$ and $p = 1/2$, the probability of earning a passing grade is

$$P(x=6) + P(x=7) + P(x=8) + P(x=9) + P(x=10)$$

$$= \binom{10}{6}\left(\frac{1}{2}\right)^6\left(\frac{1}{2}\right)^4 + \binom{10}{7}\left(\frac{1}{2}\right)^7\left(\frac{1}{2}\right)^3 + \binom{10}{8}\left(\frac{1}{2}\right)^8\left(\frac{1}{2}\right)^2$$

$$+ \binom{10}{9}\left(\frac{1}{2}\right)^9\left(\frac{1}{2}\right)^1 + \binom{10}{10}\left(\frac{1}{2}\right)^{10}\left(\frac{1}{2}\right)^0 = \frac{386}{1024} = 0.37695$$

11. a) The probability of drawing a red ball on a single draw is six chances out of ten, so $p = 0.6$. Then, $n = 10$ and $x = 6$

$$P(x=6) = \binom{10}{6}(0.6)^6(0.4)^4 = 0.2508$$

b) The probability of drawing a white ball is 1/10 and $n = 7$. To have at least 1 white ball, there could be 1, 2, 3, 4, 5, 6, or 7 white balls drawn in seven trials. The complementary event "no white balls are drawn." Therefore, $(1 - P(x=0))$ is

$$1 - \binom{7}{0}\left(\frac{1}{10}\right)^0\left(\frac{9}{10}\right)^7 = 1 - (0.9)^7 = 0.5217$$

c) The probability of drawing a blue ball is 3/10 and here $n = 4$.

$$P(x=4) = \binom{4}{4}\left(\frac{3}{10}\right)^4\left(\frac{7}{10}\right)^0 = \left(\frac{3}{4}\right)^4 = \frac{81}{10{,}000} = 0.0081$$

13.
$$\begin{aligned} P(x \geq 2) &= 1 - P(x = 0) - P(x = 1) \\ &= 1 - \binom{5}{0}(0.15)^0(0.85)^5 - \binom{5}{1}(0.15)^1(0.85)^4 \\ &= 1 - 0.4437 - 0.3915 = 0.1648 \end{aligned}$$

15. $P(x = 10) = \binom{10}{10}(0.9)^{10}(0.10)^0 = 0.3487$

17. a) $P(x \geq 5) = 1 - P(x \leq 4)$

$$\begin{aligned} &= 1 - \binom{30}{0}(0.1)^0(0.9)^{30} - \binom{30}{1}(0.1)^1(0.9)^{29} \\ &\quad - \binom{30}{2}(0.1)^2(0.9)^{28} - \binom{30}{3}(0.1)^3(0.9)^{27} - \binom{30}{4}(0.1)^4(0.9)^{26} \\ &= 1 - 0.0424 - 0.1413 - 0.2277 - 0.2361 - 0.1771 = 0.1754 \end{aligned}$$

b) $P(x = 0) = \binom{30}{0}(0.1)^0(0.9)^{30} = 0.0424$

19. The complement event has fewer cases, so one seeks $P(x = 0) \leq 0.05$

$$\begin{aligned} \binom{n}{0}\left(\frac{1}{6}\right)^0\left(\frac{5}{6}\right)^n &\leq 0.05 \\ \left(\frac{5}{6}\right)^n &\leq 0.05 \\ (n)\ln(5/6) &\leq \ln(0.05) \\ n &\geq \frac{\ln(0.05)}{\ln(5/6)} = 16.43 \end{aligned}$$

There must be at least 17 rolls of the die.

EXERCISES 8.3

1. a) For three red and 0 black (chose three of six red and 0 of four black)

$$\frac{\binom{6}{3}\binom{4}{0}}{\binom{10}{3}} = \frac{(20)(1)}{120} = \frac{1}{6}$$

⟶

b) If both colors are represented, there are either (two red and one black) or (one red and two black).

$$\frac{\binom{6}{2}\binom{4}{1}+\binom{6}{1}\binom{4}{2}}{\binom{10}{3}}=\frac{60+36}{120}=0.8$$

3. a) To get all six correct, six out of the six winning numbers were chosen and none of the 30 losing numbers. So,

$$\frac{\binom{6}{6}\binom{30}{0}}{\binom{36}{6}}=\frac{1}{1,947,792}$$

b) To get five numbers correct, five of the six winning numbers and one of the 30 losing numbers were chosen. So,

$$\frac{\binom{6}{5}\binom{30}{1}}{\binom{36}{6}}=\frac{180}{1{,}947{,}792}$$

5. a) If the hand has four aces then all of the aces were drawn and one of the other 48 cards.

$$\frac{\binom{4}{4}\binom{48}{1}}{\binom{52}{5}}=\frac{48}{2{,}598{,}960}.$$

b) A flush can be in any of the four suits. You must then pick 5 of the 13 cards of that suit and none of the other 39 cards.

$$4\left(\frac{\binom{13}{5}\binom{39}{0}}{\binom{52}{5}}\right)=4\left(\frac{1287}{2{,}598{,}960}\right)=0.00198$$

7. Think of two groups; one of three and the other of five.

a) If no minority candidate is hired, then the three hired people were from the group of five, so

$$\frac{\binom{3}{0}\binom{5}{3}}{\binom{8}{3}} = \frac{(1)(10)}{56} = \frac{5}{28}$$

b) Here, the three minority candidates are hired and none of the other five are hired. One has

$$\frac{\binom{3}{3}\binom{5}{0}}{\binom{8}{3}} = \frac{(1)(1)}{56} = \frac{1}{56}$$

9. There are two groups: the 6 defectives and the 14 "good" items.

a) "No defectives" means that 5 of the 14 "good" are chosen. So,

$$\frac{\binom{6}{0}\binom{14}{5}}{\binom{20}{5}} = \frac{2002}{15{,}504} = 0.1291$$

b) Here, chose one of the six defectives and four of the others:

$$\frac{\binom{6}{1}\binom{14}{4}}{\binom{20}{5}} = \frac{6006}{15{,}504} = 0.3874$$

c) Here, chose five of the six defectives and none of the others:

$$\frac{\binom{6}{5}\binom{14}{0}}{\binom{20}{5}} = \frac{6}{15{,}504} = 0.000387$$

11. Here, several combinations are multiplied to get the numerator

$$\frac{\binom{5}{3}\binom{4}{2}\binom{1}{0}\binom{2}{1}}{\binom{12}{6}} = \frac{(10)(6)(1)(2)}{924} = \frac{120}{924} = 0.1299$$

13.

$$\frac{\binom{53}{13}\binom{45}{12}\binom{2}{0}}{\binom{100}{25}} = \frac{2.4198 \times 10^{22}}{2.4252 \times 10^{23}} = 0.0998$$

15.

$$\frac{\binom{3}{1}\binom{6}{3}}{\binom{9}{4}} = \frac{60}{126} = \frac{10}{21}$$

(assuming that there are three women on the current Supreme Court)

EXERCISES 8.4

1. Here, $\lambda = 2.6$

a) $$P(x \le 3) = \frac{2.6^0 e^{-2.6}}{0!} + \frac{2.6^1 e^{-2.6}}{1!} + \frac{2.6^2 e^{-2.6}}{2!} + \frac{2.6^3 e^{-2.6}}{3!}$$
$$= 0.0743 + 0.1931 + 0.2510 + 0.2176 = 0.7360$$

b) $$P(x = 1) = \frac{2.6^1 e^{-2.6}}{1!} = 0.1931$$

c) $$P(x > 2) = 1 - P(x \le 2)$$
$$= 1 - \frac{2.6^0 e^{-2.6}}{0!} - \frac{2.6^1 e^{-2.6}}{1!} - \frac{2.6^2 e^{-2.6}}{2!}$$
$$= 1 - 0.0743 - 0.1931 - 0.2510 = 0.4816$$

3. a) $$P(x = 3) = \frac{2.4^3 e^{-2.4}}{3!} = 0.2090$$

b) $$P(x \le 4) = P(x = 0) + P(x = 1) + P(x = 2) + P(x = 3) + P(x = 4)$$
$$= \frac{2.4^0 e^{-2.4}}{0!} + \frac{2.4^1 e^{-2.4}}{1!} + \frac{2.4^2 e^{-2.4}}{2!} + \frac{2.4^3 e^{-2.4}}{3!} + \frac{2.4^4 e^{-2.4}}{4!}$$
$$= 0.0907 + 0.2177 + 0.2612 + 0.2090 + 0.1254 = 0.9040$$

5. The average number of errors yields $\lambda = 1$

a) $P(x = 0) = \dfrac{1^0 e^{-1}}{0!} = e^{-1} = 0.3679$

b) $$\begin{aligned} P(x \geq 3) &= 1 - P(x \leq 2) \\ &= 1 - P(x = 0) - P(x = 1) - P(x = 2) \\ &= 1 - \frac{1^0 e^{-1}}{0!} - \frac{1^1 e^{-1}}{1!} - \frac{1^2 e^{-1}}{2!} = 0.0803 \end{aligned}$$

7. Here, $\lambda = 2.6$

a) $P(x = 3) = \dfrac{2.6^3 e^{-2.6}}{3!} = 0.2176$

b) $P(x \leq 4) = P(x = 0) + P(x = 1) + P(x = 2) + P(x = 3) + P(x = 4)$

$$= \frac{2.6^0 e^{-2.6}}{0!} + \frac{2.6^1 e^{-2.6}}{1!} + \frac{2.6^2 e^{-2.6}}{2!} + \frac{2.6^3 e^{-2.6}}{3!} + \frac{2.6^4 e^{-2.6}}{4!} = 0.8774$$

c) $$\begin{aligned} P(x \geq 2) &= 1 - P(x \leq 1) \\ &= 1 - \frac{2.6^0 e^{-2.6}}{0!} - \frac{2.6^1 e^{-2.6}}{1!} = 0.7326 \end{aligned}$$

9. There are nine square feet per square yard. Therefore, $\lambda = 2(9) = 18$.

$$P(x = 0) = \frac{18^0 e^{-18}}{0!} = e^{-18} \approx 1.5 \times 10^{-8}$$

11. Using five-minute periods, $\lambda = 3$. The probability that no customers arrive is $P(x = 0) = \dfrac{(3)^0 e^{-3}}{0!} = 0.0498$. The probability that more than five customers arrive is found using the complement event $1 - P(x \leq 5)$

$$1 - \frac{3^0 e^{-3}}{0!} - \frac{3^1 e^{-3}}{1!} - \frac{3^2 e^{-3}}{2!} - \frac{3^3 e^{-3}}{3!} - \frac{3^4 e^{-3}}{4!} - \frac{3^5 e^{-3}}{5!} = 0.0839$$

13. Here, $\lambda = 2.4$. The probability of exactly four defects is $\dfrac{(2.4)^4 e^{-2.4}}{4!} = 0.1254$. The probability of less than two breakdowns is

$$\begin{aligned} P(x < 2) &= P(x = 0) + P(x = 1) \\ &= \frac{2.4^0 e^{-2.4}}{0!} + \frac{2.4^1 e^{-2.4}}{1!} = 0.3084 \end{aligned}$$

SUPPLEMENTARY EXERCISES CHAPTER 8

1. The entire probability distribution is:

x	10	20	40	60	75	100
$P(x)$	0.20	0.15	0.05	0.30	0.20	0.10

The mean is

$$E(x) = 10(0.2) + 20(0.15) + 40(0.15) + 60(0.30) + 75(0.2) + 100(0.10) = 50$$

3. The probability of a girl is 1/2 as is the probability of a boy.

a) At least two girls are

$$\begin{aligned} P(x \geq 2) &= P(x = 2) + P(x = 3) + P(x = 4) + P(x = 5) \\ &= \binom{5}{2}\left(\frac{1}{2}\right)^2\left(\frac{1}{2}\right)^3 + \binom{5}{3}\left(\frac{1}{2}\right)^3\left(\frac{1}{2}\right)^2 \\ &\quad + \binom{5}{4}\left(\frac{1}{2}\right)^4\left(\frac{1}{2}\right)^1 + \binom{5}{5}\left(\frac{1}{2}\right)^5\left(\frac{1}{2}\right)^0 \\ &= \frac{10}{32} + \frac{10}{32} + \frac{5}{32} + \frac{1}{32} = \frac{26}{32} = 0.8125 \end{aligned}$$

b) If all children are of like gender, one seeks

$$\begin{aligned} P(x = 0) + P(x = 5) &= \binom{5}{0}\left(\frac{1}{2}\right)^0\left(\frac{1}{2}\right)^5 + \binom{5}{5}\left(\frac{1}{2}\right)^5\left(\frac{1}{2}\right)^0 \\ &= \frac{1}{32} + \frac{1}{32} = 0.0625 \end{aligned}$$

5. In each case, one seeks $P(x \geq 2) = 1 - P(x \leq 1)$.

a) Here, $p = 0.05$ and $n = 10$

$$1 - \binom{10}{0}(0.05)^0(0.95)^{10} - \binom{10}{1}(0.05)^1(0.95)^9 = 0.0861$$

⟶

b) Here, $p = 0.10$ and $n = 10$

$$1 - \binom{10}{0}(0.10)^0(0.90)^{10} - \binom{10}{1}(0.10)^1(0.90)^9 = 0.2639$$

c) Here, $p = 0.25$ and $n = 10$

$$1 - \binom{10}{0}(0.25)^0(0.75)^{10} - \binom{10}{1}(0.25)^1(0.75)^9 = 0.7560$$

7. The probability distribution for this die toss experiment is:

x	1	2	3	4	5	6
$P(x)$	1/9	2/9	1/9	2/9	1/9	2/9

The expected value is $(1)(1/9) + (2)(2/9) + \cdots + (6)(2/9) = 33/9$
The player is charged \$3.67 for the game to be fair.

9. There are three defectives and 17 others. Two defectives and three of the others are to be chosen at one time. The hypergeometric distribution applies to yield

$$\frac{\binom{3}{2}\binom{17}{3}}{\binom{20}{5}} = 0.1316$$

11. The hypergeometric distribution applies with four aces and 48 other cards. The probability distribution for the number of aces is

x	0	1	2	3	4
$P(x)$	$\frac{\binom{4}{0}\binom{48}{5}}{\binom{52}{5}}$ 0.6588	$\frac{\binom{4}{1}\binom{48}{4}}{\binom{52}{5}}$ 0.2995	$\frac{\binom{4}{2}\binom{48}{3}}{\binom{52}{5}}$ 0.0399	$\frac{\binom{4}{3}\binom{48}{2}}{\binom{52}{5}}$ 0.0017	$\frac{\binom{4}{4}\binom{48}{1}}{\binom{52}{5}}$ ≈ 0

13. a) $$P(x \leq 4) = \frac{(2.4)^0 e^{-2.4}}{0!} + \frac{(2.4)^1 e^{-2.4}}{1!} + \frac{(2.4)^2 e^{-2.4}}{2!} + \frac{(2.4)^3 e^{-2.4}}{3!} + \frac{(2.4)^4 e^{-2.4}}{4!}$$
$$= 0.0907 + 0.2177 + 0.2613 + 0.2090 + 0.1254 = 0.9041$$

b) $P(x = 2) = 0.2613$

c) $P(x > 1) = 1 - P(x \leq 1) = 1 - 0.907 - 0.2177 = 0.6916$

15. a) The mean is $\frac{10 + 8 + \cdots + 12}{8} = 10$

b) $P(x = 9) = \frac{(10)^9 e^{-10}}{9!} = 0.251$

c) $P(x \geq 8) = 1 - P(x \leq 7) = 0.7798$

CHAPTER 9

MARKOV CHAINS

EXERCISES 9.1

1. Each row in a Markov chain transition probability matrix must sum to unity.

a) $0.2 + x = 1$ so $x = 0.8$, $y + 0.3 = 1$ so $y = 0.7$

b) $0.4 + 0.5 + x = 1$ so $x = 0.1x + y + 0.3 = 1$ and using $x = 0.1$ yields $y = 0.6$

c) $0.7 + 2y + y = 1$ so $y = 0.1$, $x + x + 2y = 1$ or $2x + 2(0.1) = 1$ therefore, $x = 0.4$.

3. The transition diagram for the chain is:

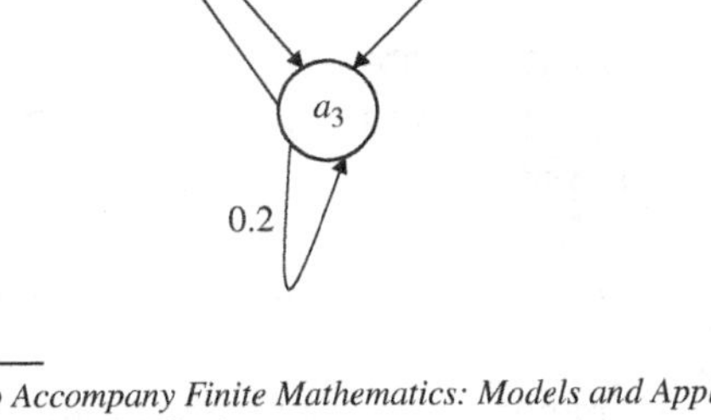

Solutions Manual to Accompany Finite Mathematics: Models and Applications, First Edition.
Carla C. Morris and Robert M. Stark.

5. a) The transition matrix for the diagram is $\begin{bmatrix} 0.85 & 0.15 \\ 0.25 & 0.75 \end{bmatrix}$.

b) The transition matrix for the diagram is $\begin{bmatrix} 0.55 & 0.45 \\ 0.60 & 0.40 \end{bmatrix}$.

7. The transition matrix for the diagram is $\begin{bmatrix} 0.7 & 0.1 & 0.0 & 0.2 \\ 0.0 & 0.8 & 0.0 & 0.2 \\ 0.6 & 0.0 & 0.3 & 0.1 \\ 0.0 & 0.1 & 0.0 & 0.9 \end{bmatrix}$.

9. For the probability of a move from State 3 to State 1, seek the entry in row 3 and column 1, which is 0.7.

EXERCISES 9.2

1. $T^2 = \begin{bmatrix} 0.2 & 0.8 \\ 0.9 & 0.1 \end{bmatrix} \begin{bmatrix} 0.2 & 0.8 \\ 0.9 & 0.1 \end{bmatrix} = \begin{bmatrix} 0.76 & 0.24 \\ 0.27 & 0.73 \end{bmatrix}$

3. a) $T(2) = \begin{bmatrix} 0.5 & 0.3 & 0.2 \\ 0.2 & 0.4 & 0.4 \\ 0.6 & 0.1 & 0.3 \end{bmatrix} \begin{bmatrix} 0.5 & 0.3 & 0.2 \\ 0.2 & 0.4 & 0.4 \\ 0.6 & 0.1 & 0.3 \end{bmatrix} = \begin{bmatrix} 0.43 & 0.29 & 0.28 \\ 0.42 & 0.26 & 0.32 \\ 0.50 & 0.25 & 0.25 \end{bmatrix}$

b) One seeks $t_{12}(2)$, which is 0.29

c) One seeks $t_{23}(2)$, which is 0.32.

5. a)
$$T(2) = T^2 = \begin{bmatrix} 0.2 & 0.3 & 0.5 \\ 0.4 & 0.5 & 0.1 \\ 0.7 & 0.1 & 0.2 \end{bmatrix} \begin{bmatrix} 0.2 & 0.3 & 0.5 \\ 0.4 & 0.5 & 0.1 \\ 0.7 & 0.1 & 0.2 \end{bmatrix}$$
$$= \begin{bmatrix} 0.51 & 0.26 & 0.23 \\ 0.35 & 0.38 & 0.27 \\ 0.32 & 0.28 & 0.40 \end{bmatrix}$$
$$T(3) = T(2)T = \begin{bmatrix} 0.51 & 0.26 & 0.23 \\ 0.35 & 0.38 & 0.27 \\ 0.32 & 0.28 & 0.40 \end{bmatrix} \begin{bmatrix} 0.2 & 0.3 & 0.5 \\ 0.4 & 0.5 & 0.1 \\ 0.7 & 0.1 & 0.2 \end{bmatrix}$$
$$= \begin{bmatrix} 0.367 & 0.306 & 0.327 \\ 0.411 & 0.322 & 0.267 \\ 0.456 & 0.276 & 0.268 \end{bmatrix}$$

b) The probability of moving from State 1 to State 3 in three steps is $t_{13}(3) = 0.327$.

c) The probability of moving from State 3 to State 2 in three steps is $t_{32}(3) = 0.276$.

7. Here, $T = \begin{bmatrix} 0.25 & 0.75 \\ 0.10 & 0.90 \end{bmatrix}$. To determine probabilities of accidents two years hence requires $T(2)$

$$T(2) = \begin{bmatrix} 0.25 & 0.75 \\ 0.10 & 0.90 \end{bmatrix} \begin{bmatrix} 0.25 & 0.75 \\ 0.10 & 0.90 \end{bmatrix} = \begin{bmatrix} 0.1375 & 0.8625 \\ 0.1150 & 0.8850 \end{bmatrix}$$

The probability of an accident two years hence, given an accident this year, is $t_{11}(2) = 0.1375$.

9. $T^2 = \begin{bmatrix} 0.23 & 0.59 & 0.18 \\ 0.22 & 0.60 & 0.18 \\ 0.18 & 0.63 & 0.19 \end{bmatrix}$. So, there is a 0.23 probability of an increase 2 weeks hence as determined by $t_{11}(2)$.

EXERCISES 9.3

1. One requires X_0T^3 to determine the solution.

$$T^2 = \begin{bmatrix} 0.6 & 0.4 \\ 0.2 & 0.8 \end{bmatrix} \begin{bmatrix} 0.6 & 0.4 \\ 0.2 & 0.8 \end{bmatrix} = \begin{bmatrix} 0.44 & 0.56 \\ 0.28 & 0.72 \end{bmatrix}$$

$$T^3 = T^2T = \begin{bmatrix} 0.44 & 0.56 \\ 0.28 & 0.72 \end{bmatrix} \begin{bmatrix} 0.6 & 0.4 \\ 0.2 & 0.8 \end{bmatrix} = \begin{bmatrix} 0.376 & 0.624 \\ 0.312 & 0.688 \end{bmatrix}$$

$$\text{Now, } X_0T^3 = \begin{bmatrix} 0.7 & 0.3 \end{bmatrix} \begin{bmatrix} 0.376 & 0.624 \\ 0.312 & 0.688 \end{bmatrix} = \begin{bmatrix} 0.3568 & 0.6432 \end{bmatrix}$$

3. Here, the weekly transition matrix is $T = \begin{bmatrix} 0.90 & 0.10 \\ 0.40 & 0.60 \end{bmatrix}$.

$T^2 = \begin{bmatrix} 0.85 & 0.15 \\ 0.60 & 0.40 \end{bmatrix}$ and $T^3 = \begin{bmatrix} 0.825 & 0.175 \\ 0.70 & 0.30 \end{bmatrix}$. One seeks X_0T^3 where

$X_0 = \begin{bmatrix} 0.30 & 0.70 \end{bmatrix}$. Therefore,

$$X_0T^3 = \begin{bmatrix} 0.30 & 0.70 \end{bmatrix} \begin{bmatrix} 0.825 & 0.175 \\ 0.70 & 0.30 \end{bmatrix} = \begin{bmatrix} 0.7375 & 0.2625 \end{bmatrix}$$

The store can expect a 73.75% share after 3 weeks of the advertising campaign.

5. a) $$T^2 = \begin{bmatrix} 0.7 & 0.3 & 0 \\ 0 & 0.8 & 0.2 \\ 0.5 & 0.3 & 0.2 \end{bmatrix}\begin{bmatrix} 0.7 & 0.3 & 0 \\ 0 & 0.8 & 0.2 \\ 0.5 & 0.3 & 0.2 \end{bmatrix} = \begin{bmatrix} 0.49 & 0.45 & 0.06 \\ 0.10 & 0.70 & 0.20 \\ 0.45 & 0.45 & 0.10 \end{bmatrix}$$

Regular since T^2 has no zero elements.

b) $$T^2 = \begin{bmatrix} 1 & 0 & 0 \\ 0.4 & 0.2 & 0.4 \\ 0.6 & 0.1 & 0.3 \end{bmatrix}\begin{bmatrix} 1 & 0 & 0 \\ 0.4 & 0.2 & 0.4 \\ 0.6 & 0.1 & 0.3 \end{bmatrix} = \begin{bmatrix} 1 & 0 & 0 \\ 0.72 & 0.08 & 0.20 \\ 0.82 & 0.05 & 0.13 \end{bmatrix}$$

All higher powers of T have the same first row, which contains zero elements and, therefore, it is not regular.

c) Regular since T has no zero elements.

7. The probability that a driver chosen at random has an accident next year is:

$$X_0T = \begin{bmatrix} 0.05 & 0.95 \end{bmatrix}\begin{bmatrix} 0.25 & 0.75 \\ 0.10 & 0.90 \end{bmatrix} = \begin{bmatrix} 0.1075 & 0.8925 \end{bmatrix} \text{ or } 10.75\%.$$

The probability of an accident the year after next is:

$$X_0T^2 = \begin{bmatrix} 0.05 & 0.95 \end{bmatrix}\begin{bmatrix} 0.1375 & 0.8625 \\ 0.1150 & 0.8850 \end{bmatrix} = \begin{bmatrix} 0.116125 & 0.883875 \end{bmatrix}$$

or 11.61%.

9. Probability vectors sum to unity, so

a) $x = 0.10$

b) $x = 1/3$

c) $x = 1/6$

11. Recall, to determine the stable probability vector $\boldsymbol{w}(\boldsymbol{T} - \boldsymbol{I}) = \boldsymbol{0}$,

$$T - I = \begin{bmatrix} 0.1 & 0.9 \\ 0.7 & 0.3 \end{bmatrix} - \begin{bmatrix} 1 & 0 \\ 0 & 1 \end{bmatrix} = \begin{bmatrix} -0.9 & 0.9 \\ 0.7 & -0.7 \end{bmatrix}.$$

Therefore,

$$\begin{bmatrix} w_1 & w_2 \end{bmatrix}\begin{bmatrix} -0.9 & 0.9 \\ 0.7 & -0.7 \end{bmatrix} = \begin{bmatrix} 0 & 0 \end{bmatrix}.$$

Next, recall that one of the equations is replaced by $w_1 + w_2 = 1$ (probability vector components sum to unity). The system is:

⟶

$$\begin{aligned} w_1 + \quad w_2 &= 1 \\ -0.9w_1 + 0.9w_2 &= 0 \end{aligned}$$

The stable probability vector is $\begin{bmatrix} 7/16 & 9/16 \end{bmatrix}$.

13. First,

$$\boldsymbol{T}-\boldsymbol{I}=\begin{bmatrix} 0 & 0 & 1 \\ 0.25 & 0.75 & 0 \\ 0 & 0.50 & 0.50 \end{bmatrix}-\begin{bmatrix} 1 & 0 & 0 \\ 0 & 1 & 0 \\ 0 & 0 & 1 \end{bmatrix}=\begin{bmatrix} -1 & 0 & 1 \\ 0.25 & -0.25 & 0 \\ 0 & 0.50 & -0.50 \end{bmatrix}$$

Next, recall, $\boldsymbol{w}(\boldsymbol{T}-\boldsymbol{I})=\boldsymbol{0}$. This gives:

$$\begin{bmatrix} w_1 & w_2 & w_3 \end{bmatrix}\begin{bmatrix} -1 & 0 & 1 \\ 0.25 & -0.25 & 0 \\ 0 & 0.50 & -0.50 \end{bmatrix}=\begin{bmatrix} 0 & 0 & 0 \end{bmatrix}$$

The resulting equations are:

$$\begin{aligned} -w_1 + 0.25w_2 + \qquad\qquad &= 0 \\ -0.25w_2 + 0.50w_3 &= 0 \\ w_1 \qquad\qquad - 0.50w_3 &= 0 \end{aligned}$$

One of the equations is replaced with $w_1 + w_2 + w_3 = 1$. Assuming that the first equation is replaced, solve the system

$$\begin{aligned} w_1 + \quad w_2 + \quad w_3 &= 1 \\ -0.25w_2 + 0.50w_3 &= 0 \\ w_1 \qquad\qquad - 0.50w_3 &= 0 \end{aligned}$$

Because the second and third equations each only have two of the three w_i's, it would be easier to solve the system by substitution.
From the second equation, $0.50w_3 = 0.25w_2$ or $w_2 = 2w_3$.
From the third equation, $w_1 = 0.50w_3$.
Substituting in the first equation yields that $0.50w_3 + 2w_3 + w_3 = 1$ or that $w_3 = 2/7$ and $w_1 = 1/7$ and $w_2 = 4/7$.

The steady-state vector is $\begin{bmatrix} 1/7 & 4/7 & 2/7 \end{bmatrix}$.

15. First, determine $\boldsymbol{T}-\boldsymbol{I}$.

$$\boldsymbol{T}-\boldsymbol{I}=\begin{bmatrix}1/8 & 3/4 & 1/8\\1/4 & 1/4 & 1/2\\1/2 & 1/4 & 1/4\end{bmatrix}-\begin{bmatrix}1 & 0 & 0\\0 & 1 & 0\\0 & 0 & 1\end{bmatrix}=\begin{bmatrix}-7/8 & 3/4 & 1/8\\1/4 & -3/4 & 1/2\\1/2 & 1/4 & -3/4\end{bmatrix}$$

Next, recall, $\boldsymbol{w}(\boldsymbol{T}-\boldsymbol{I})=\boldsymbol{0}$. This gives:

$$\begin{bmatrix}w_1 & w_2 & w_3\end{bmatrix}\begin{bmatrix}-7/8 & 3/4 & 1/8\\1/4 & -3/4 & 1/2\\1/2 & 1/4 & -3/4\end{bmatrix}=\begin{bmatrix}0 & 0 & 0\end{bmatrix}$$

The resulting equations are:

$$-7/8w_1+1/4w_2+1/2w_3=0$$
$$3/4w_1-3/4w_2+\ 1/4w_3=0$$
$$1/8w_1+1/2w_2\ -3/4w_3=0$$

One of the redundant equations is replaced with $w_1+w_2+w_3=1$. Assuming that the first equation is replaced, solve

$$w_1+\quad w_2+\quad w_3=1$$
$$3/4w_1-3/4w_2+1/4w_3=0$$
$$1/8w_1+1/2w_2-3/4w_3=0$$

Solving, the steady-state vector is $\begin{bmatrix}14/48 & 19/48 & 15/48\end{bmatrix}$.

17. Answers vary with the state data.
For Delaware, one has:
Using D to denote Democrat and R to denote Republican: There are four elections where a Democrat was reelected DD, there are four DR transitions, five RD transitions, and 12 RR transitions. Recall, the row probabilities sum to unity. The transition matrix for the election results is

$$\begin{array}{cc} & \begin{array}{cc}D & \quad R\end{array}\\ \begin{array}{c}D\\R\end{array} & \begin{bmatrix}1/2 & 1/2\\5/17 & 12/17\end{bmatrix}\end{array}$$

Solving $\boldsymbol{w}(\boldsymbol{T}-\boldsymbol{I})=\boldsymbol{0}$ yields a steady-state vector $\begin{bmatrix}10/27 & 17/27\end{bmatrix}$.

EXERCISES 9.4

1. a) State 1 is an absorbing state since there is a 1 on the main diagonal. From nonabsorbing State 2, it is possible to reach State 1, so this is an absorbing Markov chain.

b) There is no absorbing state, so it is not an absorbing Markov chain.

c) State 3 is the only absorbing state. It is possible to reach State 3 directly from nonabsorbing State 1. The same for nonabsorbing State 4. From nonabsorbing State 2, it is possible to reach the absorbing state via State 1. Therefore, because the absorbing state can be reached from all the nonabsorbing states, it is an absorbing Markov chain.

d) Here, there are two absorbing States 2 and 3. Direct movement from State 1 to State 3 is possible. Direct movement from State 4 to State 2 is possible. Movement from State 5 to State 3 is possible via State 1. Since an absorbing state can be reached by all of the nonabsorbing states, this is an absorbing Markov chain.

3. State 3 is an absorbing state since it has an arrow leaving and reentering it with a probability of 1. State 3 can be reached directly from both nonabsorbing States 1 and 2, so it is an absorbing Markov chain.

5. a) State 2 is an absorbing state. Reordering the States 2, 1, 3, 4, the canonical form is as follows.

$$\begin{array}{c} \\ 2 \\ 1 \\ 3 \\ 4 \end{array}\begin{array}{c} \begin{array}{cccc} 2 & 1 & 3 & 4 \end{array} \\ \left[\begin{array}{c|ccc} 1.0 & 0.0 & 0.0 & 0.0 \\ \hline 0.4 & 0.2 & 0.2 & 0.2 \\ 0.0 & 0.5 & 0.3 & 0.2 \\ 0.6 & 0.0 & 0.0 & 0.4 \end{array}\right] \end{array}$$

b) $I - S$ is:

$$\begin{bmatrix} 1 & 0 & 0 \\ 0 & 1 & 0 \\ 0 & 0 & 1 \end{bmatrix} - \begin{bmatrix} 0.2 & 0.2 & 0.2 \\ 0.5 & 0.3 & 0.2 \\ 0.0 & 0.0 & 0.4 \end{bmatrix} = \begin{bmatrix} 0.8 & -0.2 & -0.2 \\ -0.50 & 0.7 & -0.2 \\ 0.0 & 0.6 & 0.6 \end{bmatrix}$$

and

$$N = (I - S)^{-1} = \begin{bmatrix} \frac{35}{23} & \frac{10}{23} & \frac{15}{23} \\ \frac{25}{23} & \frac{40}{23} & \frac{65}{69} \\ 0 & 0 & \frac{5}{3} \end{bmatrix}$$

7. Traps are in Compartments 2 and 5, so these are the absorbing states. Relabeling states as 2, 5, 1, 3, 4 gives the canonical form:

$$\left[\begin{array}{cc|ccc} 1 & 0 & 0 & 0 & 0 \\ 0 & 1 & 0 & 0 & 0 \\ \hline 1/2 & 0 & 0 & 0 & 1/2 \\ 1/3 & 1/3 & 0 & 0 & 1/3 \\ 0 & 1/3 & 1/3 & 1/3 & 0 \end{array}\right]$$

$$\boldsymbol{I}-\boldsymbol{S}=\begin{bmatrix} 1 & 0 & 0 \\ 0 & 1 & 0 \\ 0 & 0 & 1 \end{bmatrix}-\begin{bmatrix} 0 & 0 & 1/2 \\ 0 & 0 & 1/3 \\ 1/3 & 1/3 & 0 \end{bmatrix}=\begin{bmatrix} 1 & 0 & -1/2 \\ 0 & 1 & -1/3 \\ -1/3 & -1/3 & 1 \end{bmatrix}$$

Finding an inverse using the method in Chapter 3, one has

$$\left[\begin{array}{ccc|ccc} 1 & 0 & -1/2 & 1 & 0 & 0 \\ 0 & 1 & -1/3 & 0 & 1 & 0 \\ -1/3 & -1/3 & 1 & 0 & 0 & 1 \end{array}\right] \to \left[\begin{array}{ccc|ccc} 1 & 0 & -1/2 & 1 & 0 & 0 \\ 0 & 1 & -1/3 & 0 & 1 & 0 \\ 0 & -1/3 & 5/6 & 1/3 & 0 & 1 \end{array}\right]$$

$$\to \left[\begin{array}{ccc|ccc} 1 & 0 & -1/2 & 1 & 0 & 0 \\ 0 & 1 & -1/3 & 0 & 1 & 0 \\ 0 & 0 & 13/8 & 1/3 & 1/3 & 1 \end{array}\right] \to \left[\begin{array}{ccc|ccc} 1 & 0 & -1/2 & 1 & 0 & 0 \\ 0 & 1 & -1/3 & 0 & 1 & 0 \\ 0 & 0 & 1 & 6/13 & 6/13 & 18/13 \end{array}\right]$$

$$\to \left[\begin{array}{ccc|ccc} 1 & 0 & 0 & 16/13 & 3/13 & 9/13 \\ 0 & 1 & 0 & 2/13 & 15/13 & 6/13 \\ 0 & 0 & 1 & 6/13 & 6/13 & 18/13 \end{array}\right] \text{ and so}$$

$$N=\begin{bmatrix} 16/13 & 3/13 & 9/13 \\ 2/13 & 15/13 & 6/13 \\ 6/13 & 6/13 & 18/13 \end{bmatrix}$$

The states for the rows and columns of N are labeled 1, 3, and 4, so one seeks N_{12}. The expected number of times the rat is in Compartment 3 before being trapped when it started in Compartment 1 is 3/13.
To answer the second part, one needs to find $\boldsymbol{C}=\boldsymbol{N}\boldsymbol{R}$

$$\boldsymbol{C}=\begin{bmatrix} 16/13 & 3/13 & 9/13 \\ 2/13 & 15/13 & 6/13 \\ 6/13 & 6/13 & 18/13 \end{bmatrix}\begin{bmatrix} 1/2 & 0 \\ 1/3 & 1/3 \\ 0 & 1/3 \end{bmatrix}=\begin{bmatrix} 9/13 & 4/13 \\ 6/13 & 7/13 \\ 5/13 & 8/13 \end{bmatrix}$$

Here, the rows are labeled 1, 3, 4 and the columns 2, 5.
One seeks $\boldsymbol{C}_{12}$. The probability if the rat starts in Compartment 1 of being trapped in Compartment 5 is 4/13.

SUPPLEMENTARY EXERCISES

1. The transition diagram is:

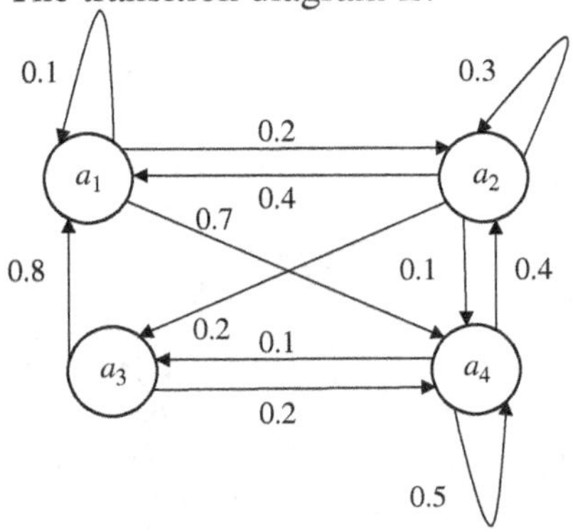

3. a) To find the three step transition matrix $\boldsymbol{T}(\mathbf{3})$, first, find $\boldsymbol{T}(\mathbf{2})$

$$\boldsymbol{T}(\mathbf{2}) = \begin{bmatrix} 0.1 & 0.2 & 0.7 \\ 0.4 & 0.5 & 0.1 \\ 0.8 & 0.0 & 0.2 \end{bmatrix} \begin{bmatrix} 0.1 & 0.2 & 0.7 \\ 0.4 & 0.5 & 0.1 \\ 0.8 & 0.0 & 0.2 \end{bmatrix} = \begin{bmatrix} 0.65 & 0.12 & 0.23 \\ 0.32 & 0.33 & 0.35 \\ 0.24 & 0.16 & 0.60 \end{bmatrix}$$

Next,

$$\boldsymbol{T}(\mathbf{3}) = \boldsymbol{T}(\mathbf{2})\boldsymbol{T} = \begin{bmatrix} 0.65 & 0.12 & 0.23 \\ 0.32 & 0.33 & 0.35 \\ 0.24 & 0.16 & 0.60 \end{bmatrix} \begin{bmatrix} 0.1 & 0.2 & 0.7 \\ 0.4 & 0.5 & 0.1 \\ 0.8 & 0.0 & 0.2 \end{bmatrix}$$

$$= \begin{bmatrix} 0.297 & 0.190 & 0.513 \\ 0.444 & 0.229 & 0.327 \\ 0.568 & 0.128 & 0.304 \end{bmatrix}$$

b) The probability that a system in State 2 is in State 1 three observations later is $\boldsymbol{t}_{\mathbf{21}}(\mathbf{3}) = 0.444$.

5. If a transition matrix is regular, the $\boldsymbol{T}$ or some power of $\boldsymbol{T}$ must have all nonzero entries.

a) This matrix is regular since $\boldsymbol{T}$ has all nonzero entries.

b) $\boldsymbol{T} = \begin{bmatrix} 1.0 & 0.0 \\ 0.5 & 0.5 \end{bmatrix}$ has an entry that is zero, so $\boldsymbol{T}^2$ is investigated next.

$\boldsymbol{T}^2 = \begin{bmatrix} 1.0 & 0.0 \\ 0.5 & 0.5 \end{bmatrix} \begin{bmatrix} 1.0 & 0.0 \\ 0.5 & 0.5 \end{bmatrix} = \begin{bmatrix} 1.0 & 0.0 \\ 3/4 & 1/4 \end{bmatrix}$, which also has a zero entry and investigating higher powers of $\boldsymbol{T}$, one

determines $\boldsymbol{t_{12}(1) = t_{12}(2) = t_{12}(3) = \cdots = t_{12}(n) = 0}$.
Therefore, this transition matrix is not regular.

c) The transition matrix $\boldsymbol{T}$ has at least one entry that is zero as does $\boldsymbol{T}^2$.
However, $\boldsymbol{T}^3$ has all nonzero entries, so this matrix is regular.

$$T^2 = \begin{bmatrix} 1/4 & 3/4 & 0 \\ 0 & 0 & 1 \\ 1/2 & 0 & 1/2 \end{bmatrix} \begin{bmatrix} 1/4 & 3/4 & 0 \\ 0 & 0 & 1 \\ 1/2 & 0 & 1/2 \end{bmatrix} = \begin{bmatrix} 1/16 & 3/16 & 3/4 \\ 1/2 & 0 & 1/2 \\ 3/8 & 3/8 & 1/4 \end{bmatrix}$$

$$T^3 = \begin{bmatrix} 1/16 & 3/16 & 3/4 \\ 1/2 & 0 & 1/2 \\ 3/8 & 3/8 & 1/4 \end{bmatrix} \begin{bmatrix} 1/4 & 3/4 & 0 \\ 0 & 0 & 1 \\ 1/2 & 0 & 1/2 \end{bmatrix} = \begin{bmatrix} 25/64 & 3/64 & 9/16 \\ 3/8 & 3/8 & 1/4 \\ 7/32 & 9/32 & 1/2 \end{bmatrix}$$

7. Using $\boldsymbol{w(T - I) = 0}$ yields

$$\begin{bmatrix} w_1 & w_2 & w_3 \end{bmatrix} \begin{bmatrix} -0.5 & 0 & 1/2 \\ 0 & -0.5 & 0.5 \\ 0.75 & 0 & -0.75 \end{bmatrix} = \begin{bmatrix} 0 & 0 & 0 \end{bmatrix}$$

Replacing one of the equations by $w_1 + w_2 + w_3 = 1$ and solving yield $w_1 = w_2 = w_3 = 1/3$. The stable probability vector is $\begin{bmatrix} 3/5 & 0 & 2/5 \end{bmatrix}$.

9. Using $\boldsymbol{w(T - I) = 0}$ yields

$$\begin{bmatrix} w_1 & w_2 & w_3 \end{bmatrix} \begin{bmatrix} b-1 & a & a \\ a & b-1 & a \\ a & a & b-1 \end{bmatrix} = \begin{bmatrix} 0 & 0 & 0 \end{bmatrix}$$

Replacing one of the equations by $w_1 + w_2 + w_3 = 1$ and solving yields $w_1 = w_2 = w_3 = \dfrac{a}{a - b + 1}$. But since the w's must sum to unity, this implies that the stable probability vector is $\begin{bmatrix} 1/3 & 1/3 & 1/3 \end{bmatrix}$.

11. There is an absorbing state. State 2 is absorbing since there is a "1" on the main diagonal. State 2 can be reached from State 1 via State 3. State 2 can be directly reached from both States 3 and 4. State 2 can be reached from State 5 via State 4. Since an absorbing state can be reached from each of the nonabsorbing states, this is a matrix for an absorbing Markov chain. Relabeling the States 2, 1, 3, 4, 5, the canonical form is:

⟶

$$\begin{array}{c} \\ 2 \\ 1 \\ 3 \\ 4 \\ 5 \end{array}\begin{array}{c}
\begin{array}{ccccc} 2 & 1 & 3 & 4 & 5 \end{array} \\
\left[\begin{array}{c|cccc}
1 & 0 & 0 & 0 & 0 \\ \hline
0 & 1/3 & 1/3 & 0 & 1/3 \\
1/4 & 1/4 & 1/2 & 0 & 0 \\
0.3 & 0.1 & 0.4 & 0.2 & 0 \\
0 & 0 & 0 & 1 & 0
\end{array}\right]\end{array}$$

13. Since Compartments #3 and #5 are absorbing states, one relabels the states in the order #3, #5, #1, #2, #4 to better track of the rat's possible movements and find the transition matrix in canonical form.

$$\begin{array}{c} \\ 3 \\ 5 \\ 1 \\ 2 \\ 4 \end{array}\begin{array}{c}
\begin{array}{ccccc} 3 & 5 & 1 & 2 & 4 \end{array} \\
\left[\begin{array}{cc|ccc}
1 & 0 & 0 & 0 & 0 \\
0 & 1 & 0 & 0 & 0 \\ \hline
0 & 0 & 0 & 1/2 & 1/2 \\
1/2 & 0 & 1/2 & 0 & 0 \\
1/3 & 1/3 & 1/3 & 0 & 0
\end{array}\right]\end{array}$$

$$I-S=\begin{bmatrix} 1 & 0 & 0 \\ 0 & 1 & 0 \\ 0 & 0 & 1 \end{bmatrix}-\begin{bmatrix} 0 & 1/2 & 1/2 \\ 1/2 & 0 & 0 \\ 1/3 & 0 & 0 \end{bmatrix}=\begin{bmatrix} 1 & -1/2 & -1/2 \\ -1/2 & 1 & 0 \\ -1/3 & 0 & 1 \end{bmatrix}$$

$$(I-S)^{-1}=\begin{bmatrix} 12/7 & 6/7 & 6/7 \\ 6/7 & 10/7 & 3/7 \\ 4/7 & 2/7 & 9/7 \end{bmatrix}$$

a) The rat started in Compartment #2, represented by the second row of N. The expected number of times the rat enters Compartment #4 before being trapped is 3/7 as denoted by the second row and third column of N.

b) The probability that the rat will eventually be trapped in Compartment #3 is the entry in the second row and first column of $C=NR=\begin{bmatrix} 5/7 & 2/7 \\ 6/7 & 1/7 \\ 4/7 & 3/7 \end{bmatrix}$. There is a 6/7 probability that the rat is trapped in Compartment #3.

CHAPTER 10

MATHEMATICAL STATISTICS

EXERCISES 10.1

1. a) "All adult residents of the community" is the population of interest.
b) The sample is the 2000 adults in the community that were polled.
c) The variable of interest is the opinion on the smoking ban.

3. a) Quantitative b) Quantitative c) Qualitative d) Qualitative

5. a) Quantitative and ratio scale
b) Quantitative and ratio scale
c) Qualitative and nominal scale
d) Qualitative and ordinal scale
e) Quantitative and interval scale

Solutions Manual to Accompany Finite Mathematics: Models and Applications, First Edition.
Carla C. Morris and Robert M. Stark.

7. The frequency bar chart is:

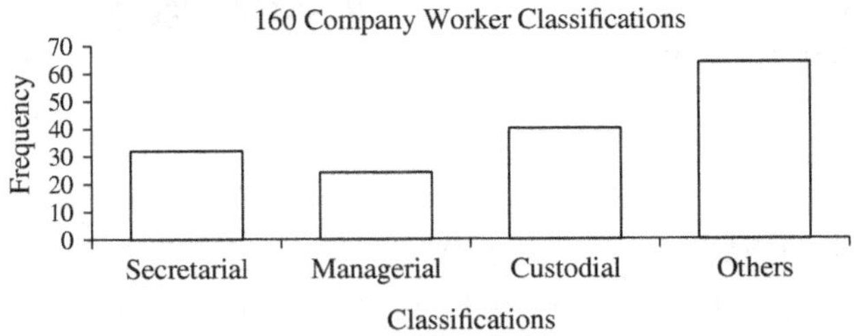

9. The pie chart is:

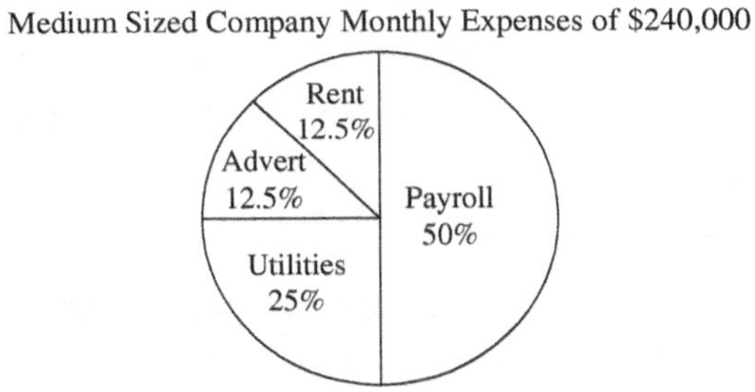

11. The frequency histogram is:

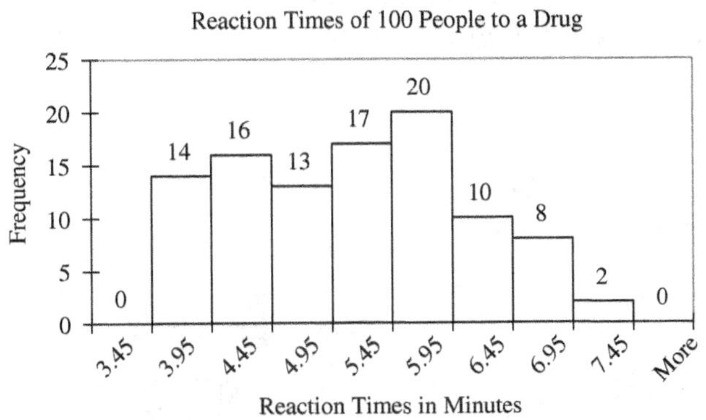

13. a) The histogram is:

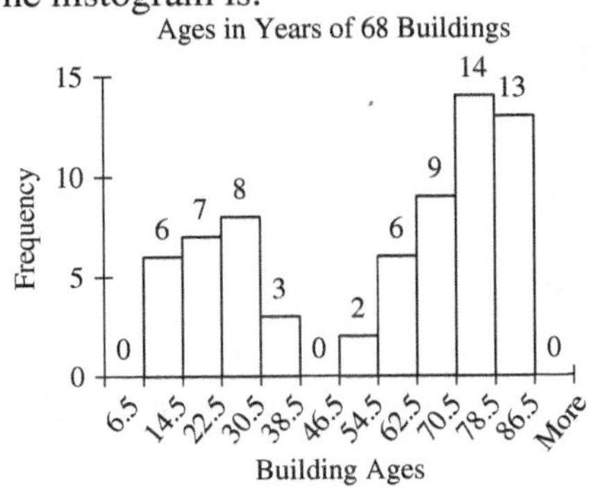

→

b) Using the graph, there are $6 + 7 + 8 + 3 + 2 + 6 + 9 = 41$ buildings less than 71 years old. The percentage is, $41/68(100\%) = 60.29\%$.

EXERCISES 10.2

1. a) $5 + 10 + 25 + 30 + 30 = 100$
   b) The summation is evaluated first. Next, subtract 5 for $100 - 5 = 95$.
   c) Subtracting 4 from each observation yields 80.
   d) Adding the squares, $25 + 1000 + 625 + 900 + 900 = 2550$.
   e) First, subtract 10 from each observation. Then find its square. Adding these yields 650.

3. The mean is $\dfrac{20 + 30 + \cdots + 50}{10} = \dfrac{390}{10} = 39$

   There are 10 observations, so the median is the average of the fifth and sixth ranked observations, in this case both are 40 and the median is 40.

   The observation occurring most often is 40.

5. There are 15 observations, so the median is the eighth ranked observation.

   The median is 381.

7. Recalling that $\bar{x} = \dfrac{\sum_{i=1}^{n} x_i}{n}$

   a) $\bar{x} = \dfrac{200}{10} = 20$

   b) $\bar{x} = \dfrac{4040}{4} = 1010$

   c) $\bar{x} = \dfrac{20}{50} = \dfrac{2}{5}$

   d) $\bar{x} = \dfrac{700}{25} = 28$

9. a) The range is $70 - 10 = 60$

   The variance is

   $$\frac{(20 - 39)^2 + (30 - 39)^2 + \cdots + (50 - 39)^2}{9} = \frac{2890}{9}$$

   The standard deviation is $\sqrt{\dfrac{2890}{9}} = \sqrt{321.1} = 17.9$

   b) The range is $11 - 0 = 11$

   The variance is $\dfrac{(8 - 6)^2 + (5 - 6)^2 + \cdots + (11 - 6)^2}{10} = \dfrac{140}{10} = 14$

   The standard deviation is $\sqrt{14} = 3.7$

11. The mean is $\frac{15 + 13 + \cdots + 23}{15} = \frac{405}{15} = 27$. The median, the eighth ranked observation, is 23. The modes are 13, 19, and 23 as they each occur twice (most frequent). The range, being the difference in the minimum and maximum values, is $67 - 13 = 54$. The variance is

$$\frac{(15-27)^2 + (13-27)^2 + \cdots + (23-27)^2}{14} = \frac{3110}{14} = 222.1$$

and the standard deviation is $\sqrt{222.1} = 14.9$.

13. a) The mean is 7.7875 and the standard deviation is 1.596

b) $\overline{x} \pm s = (6.19, 9.38)$. There are 26 observations in this interval.
$\overline{x} \pm 2s = (4.60, 10.98)$. There are 39 observations in this interval.
$\overline{x} \pm 3s = (3.00, 12.58)$. There are 39 observations in this interval.
Yes, it seems to roughly follow the empirical rule except possibly for the 13.6 measurement.

EXERCISES 10.3

1. A uniform distribution for $c = 10$ to $d = 25$ has $f(x) = \frac{1}{25-10} = \frac{1}{15}$

The mean for this uniform distribution is $\frac{10+25}{2} = 17.5$.

The standard deviation is $\frac{25-10}{\sqrt{12}} = 4.330$ and the variance is 18.75.

3. It may be useful to sketch the distribution. The interval from 15 to 30 has unit probability.

a) $P(x > 17) = \frac{30-17}{15} = \frac{13}{15}$

b) $P(x - 19) = 0$ (individual values have zero probability for continuous distributions, such as the uniform distribution).

c) $P(x \geq 12) = \frac{30-15}{15} = 1$ (there is zero probability for $x < 15$)

d) $P(x < 27.5) = \frac{27.5-15}{15} = \frac{12.5}{15} = 0.83$

e) $P(20 \leq x \leq 25) = \frac{25-20}{15} = \frac{5}{15} = 0.33$

f) $P(23 \leq x \leq 33) = \frac{30-23}{15} = \frac{7}{15} = 0.47$ (recall that there is zero probability for values of $d \geq 30$).

5. Here, $c = 100$ and $d = 150$, so this interval is the sample space.

a) $\frac{150 - a}{50} = 0.3, \ \ 150 - a = 15, \ \ a = 135$

b) $\frac{a - 100}{50} = 0.10, \ \ a - 100 = 5, \ \ a = 105$

c) $\frac{140 - a}{50} = 0.2, \ \ 140 - a = 10, \ \ a = 130$

d) $\frac{a - 100}{50} = 0.80, \ \ a - 100 = 40, \ \ a = 140$

e) $\frac{a - 105}{50} = 0.40, \ \ a - 105 = 20, \ \ a = 125$

f) $\frac{a - 128}{50} = 0.70 \ \ a - 128 = 35, \ \ a = 163$, this is not possible since there is zero probability for values of $d \geq 150$.

7. a) $\mu = \frac{0.25 + 0.35}{2} = 0.30$ and $\sigma = \frac{0.35 - 0.25}{\sqrt{12}} = 0.03$

b) $\frac{0.32 - 0.28}{0.10} = 0.40$

c) $\frac{a - 0.27}{0.1} = 0.4 \ \ a - 0.27 = 0.04$ and, therefore, $a = 0.31$.

EXERCISES 10.4

1. A bell curve sketch is useful. Here, the probabilities are directly from a Normal table.

a) 0.4177 b) 0.3389 c) 0.4846 d) 0.3729 e) 0.4913 f) 0.0239

3. A bell curve sketch reveals Z scores on opposite sides of zero. The two associated probabilities (table) are added for the required probabilities.

a) $0.3907 + 0.4929 = 0.8836$

b) $0.4952 + 0.3508 = 0.8460$

c) $0.0636 + 0.4970 = 0.5606$

d) $0.4962 + 0.2967 = 0.7929$

e) $0.3461 + 0.3461 = 0.6922$

f) $0.4788 + 0.5000 = 0.9788$ (use 0.5000 since Z is larger than tables usually show).

5. a) This includes the entire upper half of the distribution. Sum 0.5000 and $P(-1.45 < Z < 0) = 0.4265$ to yield 0.9265.

b) This is in the upper tail so $0.5000 - 0.4772 = 0.0228$

c) This is the lower tail so $0.5000 - 0.4131 = 0.0869$

d) This is the entire lower half of the curve plus $P\ (0 < Z < 1.50)$. Therefore, $0.5000 + 0.4332 = 0.9332$.

7. The two Z scores in each of these problems are on the same side of zero, so the two probabilities in the table are subtracted to determine the required probability.

a) $0.4929 - 0.3907 = 0.1022$

b) $0.4767 - 0.3665 = 0.1102$

c) $0.4977 - 0.3485 = 0.1492$

9. Here, $\mu = 80$ and $\sigma = 5$.

a) $Z = \dfrac{85 - 80}{5} = 1.00$ d) $Z = \dfrac{71 - 80}{5} = -1.80$

b) $Z = \dfrac{77 - 80}{5} = -0.60$ e) $Z = \dfrac{86 - 80}{5} = 1.20$

c) $Z = \dfrac{69 - 80}{5} = -2.20$ f) $Z = \dfrac{80 - 80}{5} = 0$

11. Here, $\mu = 80$ and $\sigma = \sqrt{16} = 4$.

a) $P(x > 78) = P(Z > -0.50) = 0.5000 + 0.1915 = 0.6915$

b) $P(74 < x < 84) = P(-1.50 < Z < 1.00) = 0.4772 + 0.3413 = 0.8185$

c) $P(x > 85) = P(Z > 1.25) = 0.5000 - 0.3944 = 0.1056$

13. Here, $\mu = 70$ and $\sigma = 3$.

a) $P(x < 76) = P(Z < 2) = 0.5000 + 0.4772 = 0.9772$

b) $P(61 < x < 73) = P(-3.00 < Z < 1.00) = 0.4987 + 0.3143 = 0.8400$

c) $P(67 \le x \le 77) = P(-1.00 \le Z \le 2.33) = 0.3413 + 0.4901 = 0.8314$.

15. a) The probability in the upper tail must be 0.1314. Therefore, $P(0 < Z < Z_0) = 0.3686$ indicating that $Z_0 = 1.12$.

b) There is symmetry, so use 0.4265 in the interior of the table to get $Z_0 = 1.45$.

c) There must be a 0.9000 probability to the left of Z_0, so one looks up $0.4000 (= 0.9000 - 0.5000)$ to yield $Z_0 = 1.28$.

⟶

d) First, for the given Z score, $P(-1 < Z < 0) = 0.3413$. There is a $0.4567(= 0.7980 - 0.3413)$ probability that Z exceeds 0. This gives $Z_0 = 1.71$.

17. Given $\sigma = 0.36$, seek $P(x > 6.5) = 0.0500$. For an upper tail probability of 0.05, look up 0.4500 in the table to get a Z score of 1.645. Therefore, $\frac{6.5 - \mu}{0.36} = 1.645$ and $\mu = 5.9078$ ml.

EXERCISES 10.5

1. a) The probability of a value within one standard deviation of the mean is 0.6826.

b) The probability of a value within two standard deviations of the mean is 0.9554.

c) The probability of a value within three standard deviations of the mean is 0.9974.

3. First, the Z value corresponding to a 2% probability in the extreme upper tail is 2.06. To determine the mean, use
$\frac{14 - \mu}{1} = 2.06$. Therefore, the mean content is set at 11.94.

5. If 45 minutes are available, one seeks to make $P(x < 45)$ as large as possible. This is the probability that the traveler arrives within the allotted time. For route A,
$P(x < 45) = P\left(Z_a < \frac{45 - 39}{4} = 1.5\right) = 0.9332$. For route B,
$P(x < 45) = P\left(Z_b < \frac{45 - 40}{3} = 1.67\right) = 0.9525$. If 45 minutes are available, route B is chosen with larger probability of arriving on time.

If the allotted time is reduced to 42 minutes,
$P\left(Z_a < \frac{42 - 39}{4} = 0.75\right) = 0.7734$. For route B,
$P\left(Z_b < \frac{42 - 40}{3} = 0.67\right) = 0.7486$. When only 42 minutes are allotted, it is advantageous to choose route A instead of route B.

7. Here, $\mu = 5$ and $\sigma = 2$. Determine $P(x > x_0) < 0.03$. The Z corresponding to a probability of 0.0300 in the extreme upper tail is $Z = 1.89$. The equation to solve is

$$\frac{x - 5}{2} = 1.89, \text{ which yields } x = 8.78$$

⟶

If the dam is built 8.78 feet high, there is less than a 3% chance of overflow given a mean of 5 feet and standard deviation of 2 feet.

9. Here, $\mu = 1000$ and $\sigma = 100$ and the associated Z with a 1% probability in the extreme upper tail is 2.33. Next, x is determined by

$$\frac{x - 1000}{100} = 2.33.$$

The seating capacity is 1233 seats in order to make sure that no more than 1% of demand would be unfulfilled.

11. a) To determine the confidence, seek $P(-1.96 < Z < 1.96)$.
The confidence level is $0.4750 + 0.4750 = 0.9500$.

b) To determine the confidence, seek the $P(-2.575 < Z < 2.575)$.
The confidence level is $0.4950 + 0.4950 = 0.9900$.

c) To determine the confidence, seek $P(-1.28 < Z < 1.28)$.
The confidence level is $0.3997 + 0.3997 = 0.7994$.

d) To determine the confidence, seek $P(-1.645 < Z < 1.645)$.
The confidence level is $0.4500 + 0.4500 = 0.9000$.

EXERCISES 10.6

1. These questions can be improved by allowing the surveyed person to choose from several options. Also, in some instances, a time period is unspecified.

In the past year, how often have you shopped on eBay?

a) Never b) 1–5 times c) 6–10 times d) more than 10 times

In which interval does your annual salary lie?

a) under $20,000 c) $40,000–$59,999

b) $20,000–$39,999 d) over $60,000

During the past year, how many books have you purchased online?

a) None b) 1–5 c) 6–10 d) more than 10

3. Answers vary. You may include questions about age, gender, smoking frequency, how long the person has smoked, and so on.

5. Answers vary.

SUPPLEMENTARY EXERCISES CHAPTER 10

1. a) Qualitative and nominal
b) Qualitative and ordinal
c) Qualitative and nominal
d) Quantitative and ratio

3. Answers vary.

5. The mean is 69.25, the median 72.5, the modes are 70, 80, and 100, the range 70, the variance 469, and the standard deviation 21.7.

7. a) $\frac{a-50}{30} = 0.2 \quad a = 56$
b) any number 50 or less
c) $\frac{70-a}{30} = 0.35 \quad a = 59.5$
d) any number larger than 80

9. a) $0.7357 - 0.5000 = 0.2357$ looking for this probability within the table yields $Z_0 = 0.63$.
b) Since it is symmetric, use $0.3400/2 = 0.1700$, which yields $Z_0 = 0.44$.
c) Here, Z_0 must be negative. Using $0.9099 - 0.5000 = 0.4099$ and looking within the Normal table yields $Z_0 = -1.34$.
d) The probability associated with the Z value of 1.1 is 0.3643. Using $0.8000 - 0.3643 = 0.4357$, so $Z_0 = -1.52$.

11. Here, $\sigma = 25$ and seek μ such that $P(x > 145) = 0.8413$. Using $0.8413 - 0.5000 = 0.3413$ and that Z is negative yields $Z = -1.00$. Therefore, $\frac{145-\mu}{25} = -1.00$ yields $\mu = 170$.

CHAPTER 11

ENRICHMENT IN FINITE MATHEMATICS

EXERCISES 11.1

1. The opponent's payoff matrix is the negative of the matrix given.

$$\begin{bmatrix} -6 & -3 & 1 \\ 5 & -3 & 4 \\ -3 & -2 & -1 \end{bmatrix}$$

3. The opponent's payoff matrix is the negative of the matrix given.

$$\begin{bmatrix} -1 & -3 & -6 \\ 2 & 4 & -5 \\ 1 & -3 & -1 \end{bmatrix}$$

5. The entries in row 1 are all larger than those in row 2, so it dominates that row. Eliminating row 2 yields

$$\begin{bmatrix} 5 & 3 & 5 & 4 \\ 6 & 1 & 4 & 2 \end{bmatrix}$$

⟶

Solutions Manual to Accompany Finite Mathematics: Models and Applications, First Edition.
Carla C. Morris and Robert M. Stark.

Next, column 2 dominates column 1, column 3, and column 4 because its entries are less than those in the other columns. Eliminating these columns yields $\begin{bmatrix} 3 \\ 1 \end{bmatrix}$.

Now the top row dominates leaving the (1, 2) entry of three a saddle point of the game.

7. Row 4 dominates row 3, since its entries are larger. Eliminating row 3 yields

$$\begin{bmatrix} 8 & 2 & 4 & 8 & 6 \\ 1 & 4 & 5 & 9 & 4 \\ 9 & 5 & 4 & 7 & 5 \end{bmatrix}$$

Now, column 2 dominates rows 4 and 5, since its entries are smaller yielding

$$\begin{bmatrix} 8 & 2 & 4 \\ 1 & 4 & 5 \\ 9 & 5 & 4 \end{bmatrix}$$

Now the third row dominates the first to yield

$$\begin{bmatrix} 1 & 4 & 5 \\ 9 & 5 & 4 \end{bmatrix}$$

9. The minimax strategy is determined by placing a circle around each row minimum and a box around each column maximum. If there are any entries with both a circle and box, they are saddle points and the game is strictly determined.

$$\begin{bmatrix} \boxed{\textcircled{2}} & 3 & \boxed{3} & \boxed{4} \\ 1 & \boxed{4} & \textcircled{-2} & 2 \\ \boxed{2} & \textcircled{1} & \boxed{3} & \boxed{4} \\ 0 & 3 & \textcircled{-3} & 1 \end{bmatrix}$$

Here, the (1, 1) entry is the saddle point of the game and the optimal strategy using the minimax criterion.

11. If the game has no saddle point, then it is nonstrictly determined. Using the circles and boxes for the minimax criterion, there is no overlap and so no saddle point.

$$\begin{bmatrix} \boxed{5} & \textcircled{2} & \boxed{3} & \boxed{4} \\ 3 & \textcircled{1} & 2 & 3 \\ 4 & 2 & \boxed{3} & \textcircled{1} \\ \boxed{5} & \boxed{3} & \boxed{3} & \textcircled{2} \end{bmatrix}$$

13. The game is solved using the LP (after 3 is added to each of the entries to remove the negatives):

Maximize: $Z = \frac{1}{v} = Y_1 + Y_2$

Subject to: $6Y_1 + 5Y_2 \leq 1$

$1Y_1 + 6Y_2 \leq 1$

$Y_1,\ Y_2 \geq 0$

Row	Y_1	Y_2	s_1	s_2	RHS
1	1	5/6	1/6	0	1/6
2	0	31/6	$-1/6$	1	5/6
Objective	0	$-1/6$	1/6	0	$1/6 = -Z$

Row	Y_1	Y_2	s_1	s_2	RHS
1	1	0	6/31	$-5/31$	1/31
2	0	1	$-1/31$	6/31	5/31
Objective	0	0	5/31	1/31	$6/31 = -Z$

$-Z = 1/v = 31/6$ so
$y_1 = (31/6)(1/31) = 1/6$, $y_2 = (31/6)(5/31) = 5/6$, from the objective row $x_1 = (31/6)(5/31) = 5/6$ and $x_2 = (31/6)(1/31) = 1/6$.

EXERCISES 11.2

1. The stock gains 100% one week and loses 50% the next. The arithmetic average is $\frac{(100\%) + (-50\%)}{2} = 25\%$. However, numerically, the \$1000 becomes \$2000 after the first week. After the second week, it is back to \$1000. Continuing, after 51 weeks it is worth \$2000 and at the end of the year \$1000, or a 0% change. Using the geometric average, weekly return gives this result since
$\sqrt{(2)(0.5)(2)(0.5)\cdots(2)(0.5)} - 1 = 0.$

3. Using the formulas from Chapter 2, the monthly payment for the car is

$$\frac{25{,}000(0.005)}{1-\dfrac{1}{(1.005)^{72}}} = \$414.32$$

5. Solution requires identifying and setting missing values.

7. The LP is

Maximize:

$$Z = (25{,}000)(1.01)^{-3}X_1 + (60{,}000)(1.01)^{-12}X_2 + (40{,}000)(1.01)^{-12}X_3 + (25{,}000)(1.01)^{-15}X_4$$

Subject to:

$$X_2 - X_3 \leq 0$$

$$X_1 \leq 3$$

$$X_4 \leq 3$$

$$X_1 \geq 1$$

$$X_2 \geq 1$$

$$X_4 \geq 1$$

$$25{,}000X_1 + 60{,}000X_2 + 40{,}000X_3 + 25{,}000X_4 = 300{,}000$$

Solving the associated linear program yields
$X_1 = 3, X_2 = 1, X_3 = 3.5, X_4 = 1$, and $Z^* = \$271{,}817.80$

9. The LP is

Maximize:

$$Z = (25{,}000)(1.01)^{-3}X_1 + (50{,}000)(1.01)^{-12}X_2 + (50{,}000)(1.01)^{-12}X_3 + (25{,}000)(1.01)^{-15}X_4$$

Subject to:

$$X_2 - X_3 \leq 0$$

$$25{,}000X_1 + 50{,}000X_2 + 50{,}000X_3 + 25{,}000X_4 = 300{,}000$$

Solving the associated linear program yields
$X_1 = 12, X_2 = 0, X_3 = 0, X_4 = 0$, and $Z^* = \$291{,}177.04$

EXERCISES 11.3

1. Recall, $SM^tP = \begin{bmatrix} 0 & 0 & 1 \end{bmatrix}$, Clan C for the sister's son who becomes the father of the grandson. That is,

$$(SM^tP)(P) = \begin{bmatrix} 0 & 0 & 1 \end{bmatrix} \begin{bmatrix} 0 & 1 & 0 \\ 0 & 0 & 1 \\ 1 & 0 & 0 \end{bmatrix} = \begin{bmatrix} 1 & 0 & 0 \end{bmatrix}$$

Therefore, the grandson is in Clan A.

The great grandson (whose father is the grandson in Clan A) is

$$\begin{bmatrix} 1 & 0 & 0 \end{bmatrix} \begin{bmatrix} 0 & 1 & 0 \\ 0 & 0 & 1 \\ 1 & 0 & 0 \end{bmatrix} = \begin{bmatrix} 0 & 1 & 0 \end{bmatrix}$$

Therefore, the great grandson is in Clan B.

3. $P = \begin{bmatrix} 0 & 1 & 0 \\ 0 & 0 & 1 \\ 1 & 0 & 0 \end{bmatrix}$, $P^2 = \begin{bmatrix} 0 & 0 & 1 \\ 1 & 0 & 0 \\ 0 & 1 & 0 \end{bmatrix} = P^3 = \begin{bmatrix} 1 & 0 & 0 \\ 0 & 1 & 0 \\ 0 & 0 & 1 \end{bmatrix} = I.$

5. Answers vary.

7. Answers vary. One possibility is

$$\begin{aligned} \text{Maximize:} \quad & 3x_{11} + x_{12} + 2x_{13} + 5x_{14} + 4x_{21} + 6x_{22} + x_{23} + 2x_{24} \\ & + 4x_{31} + 5x_{32} + 5x_{33} + x_{34} + 3x_{41} + 2x_{42} + 5x_{43} + 6x_{44} \\ \text{Subject to:} \quad & x_{11} + x_{12} + x_{13} + x_{14} \le 1 \\ & x_{21} + x_{22} + x_{23} + x_{24} \le 1 \\ & x_{31} + x_{32} + x_{33} + x_{34} \le 1 \\ & x_{41} + x_{42} + x_{43} + x_{44} \le 1 \\ & x_{11} + x_{21} + x_{31} + x_{41} \le 1 \\ & x_{12} + x_{22} + x_{32} + x_{42} \le 1 \\ & x_{13} + x_{23} + x_{33} + x_{43} \le 1 \\ & x_{14} + x_{24} + x_{34} + x_{44} \le 1 \\ & 0 \le x_{ij} \le 1 \end{aligned}$$

⟶

The objective maximum is 20, $x_{11}=1$, $x_{22}=1$, $x_{33}=1$, $x_{44}=1$, all other $x_{ij}=0$. So, each man is assigned to one woman and each woman to one man for four "monogamous" couples.

9. The observer may have been considering individuals who find no happiness in the model when the numbers of men and women are unequal.

EXERCISES 11.4

1. There are many ways to arrange the Monte Carlo. One example is to keep track of one player's winnings. A head results in Jack paying Jill \$1 dollar (+\$1). A tail results in Jill paying Jack a dollar (−\$1).
Use a column chosen randomly, say column 5, from the random number table in Chapter 6 of the text. Use an odd number to represent "a head" and an even number "a tail," so tosses can be simulated and results tabulated.

FIRST TEN TOSSES

Random Number	Result of toss	Jack's Winnings
0428	T	+1
1139	H	0
6965	H	−1
9854	T	0
6326	T	+1
3123	H	0
0857	H	−1
9972	T	0
1981	H	−1
3984	T	0

⟶

SECOND TEN TOSSES

Random Number	Result of toss	Jack's Winnings
7297	H	−1
1321	H	−1
3772	T	−1
5671	H	−1
5834	T	−1
3381	H	−1
2228	T	−1
2795	H	−1
8974	T	−1
1903	H	−2

After 10 simulated tosses, the result is \$0 and after 20 it was -\$2. Thus, in 10 tosses, the two are even but after 20 tosses Jill is ahead \$2 or an average of 10 cents per toss. In reality, the expected gain for Jill (or Jack) over many tosses is \$0.

3. Here, choose three-digit random numbers in sets of three to represent currencies A, B, and C. If the first random number is between 000 and 449, then A advances in the trading period; ranges from 450 to 799, currency A declines one unit; and for ranges 800 to 999 currency A remains unchanged for the trading period. Use a similar scheme for currency B with cutoffs at 399, 849, and 999 and for currency C at 399, 749, and 999. For each period, use $+1$ if the currency advances, 0 if it remains unchanged, and -1 if it declines. For each trading period, the gain/loss ranges from -3 if all decline in the period to $+3$ if the three currency types advance for the period.

Starting with the first column, the three random numbers are 943, 908, and 146. They indicate that A and B are unchanged, and C advances. This yields a table value of $+1$ for the day since there are equal shares of each currency.

After 4 weeks as shown in the following table, the expected gain/loss for the periods is $11/4 = \$2.75$ average weekly gain.
In reality, the expected gain/loss is $(0.45 + 0.40 + 0.40)(+1)$ $+ (0.35 + 0.45 + 0.35)(-1) + (0.20 + 0.15 + 0.25)(0) = 0.10/\text{day}$ or $\$0.50/\text{week}$.

→

Week #	Currency A	Currency B	Currency C		Weekly gain/loss
1	Unchanged	Unchanged	Advance	+1	
	Decline	Decline	Decline	−3	
	Advance	Advance	Unchanged	+2	
	Advance	Decline	Advanced	+1	
	Advance	Advance	Decline	+1	+2
2	Advance	Decline	Advance	+1	
	Decline	Advance	Advance	+1	
	Advance	Advance	Advance	+3	
	Decline	Advance	Advance	+1	
	Decline	Advance	Unchanged	0	+6
3	Decline	Advance	Unchanged	0	
	Advance	Decline	Unchanged	0	
	Decline	Decline	Advance	−1	
	Unchanged	Decline	Unchanged	−1	
	Advance	Advance	Advance	+3	+1
4	Unchanged	Advance	Advance	+2	
	Decline	Unchanged	Advance	0	
	Unchanged	Decline	Advance	0	
	Decline	Advance	Advance	+1	
	Decline	Advance	Decline	−1	+2

5. The center of the circle is $(-1, 4)$ and radius 3. The values of x lie within the interval $-1 - 3 = -4$ to $-1 + 3 = 2$ and the y values within the interval $4 - 3 = 1$ to $4 + 3 = 7$. Next, define $X_r = -4 + 6R_x$ and $Y_r = 1 + 6R_y$

a) $X_r = -4 + 6(0.61) = -0.34$ and $Y_r = 1 + 6(0.20) = 2.2$

$$(-0.34 + 1)^2 + (2.2 - 4)^2 \leq 9$$

$$3.68 \leq 9 \text{ yes}$$

b) $X_r = -4 + 6(0.24) = -2.56$ and $Y_r = 1 + 6(0.71) = 5.26$

$$(-2.56 + 1)^2 + (5.26 - 4)^2 \leq 9$$

$$4.02 \leq 9 \text{ yes}$$

⟶

c) $X_r = -4 + 6(0.68) = 0.08$ and $Y_r = 1 + 6(0.84) = 6.04$

$$(0.08 + 1)^2 + (6.04 - 4)^2 \leq 9$$

$$5.33 \leq 9 \text{ yes}$$

7. Here, the last column of the random number table was used for the random numbers. The center of the circle is (3, 4) and radius 10. The x values lie within the interval $3 - 10 = -7$ to $3 + 10 = 13$ and the y values within interval $4 - 10 = -6$ to $4 + 10 = 14$. Next, define $X_r = -7 + 20R_x$ and $Y_r = -6 + 20R_y$.

The following table summarizes the results of the first 5 and next 10 trials.

Trial	R_x	R_y	X_r	Y_r	$(X_r - 3)^2$	$(Y_r - 4)^2$	$(X_r - 3)^2 + (Y_r - 4)^2$	Hit?
1	0.46	0.92	2.2	12.4	0.64	70.56	71.2	Yes
2	0.62	0.33	5.4	0.6	5.76	11.56	17.32	Yes
3	0.44	0.80	2.0	10.0	1.00	36.0	37.0	Yes
4	0.86	0.25	10.2	−1.0	51.84	25.0	76.84	Yes
5	0.71	0.46	7.2	3.2	17.64	0.64	18.28	Yes
6	0.40	0.54	1.0	4.8	4.00	0.64	4.64	Yes
7	0.13	0.15	−4.4	−3.0	54.76	40.45	95.21	Yes
8	0.79	0.42	8.8	2.4	33.64	2.56	36.2	Yes
9	0.59	0.40	4.8	2.0	3.24	4.00	7.24	Yes
10	0.15	0.38	−4.0	1.6	49.0	5.76	54.76	Yes
11	0.89	0.88	10.8	11.6	60.84	57.76	118.60	No
12	0.59	0.52	4.8	4.4	3.24	0.16	3.4	Yes
13	0.73	0.34	7.6	0.8	21.16	10.24	31.4	Yes
14	0.65	0.53	6.0	4.6	9.0	0.36	9.36	Yes
15	0.09	0.98	−5.2	13.6	67.24	92.16	159.4	No

The area of a 20-by-20 square is 400 square units. In the case of the first five trials, all were hits so the estimate of the area of the circle is 400 square units when in reality it is approximately 314 square units. When 15 trials are used, the estimate is then $400(13/15) = 346.67$ square units, which is closer but still the estimate does not appear to have settled. It can take thousands of trials for a reliable estimate. This example illustrates the procedure for a Monte Carlo experiment.

9. Here, after the ordered pair (X_r, Y_r) is determined, one judges whether or not it is a hit (part of the shaded region).

⟶

Trial	R_x	R_y	X_r	Y_r	Hit?
1	0.97	0.27	4.85	1.08	Yes
2	0.27	0.12	1.35	0.48	No
3	0.05	0.72	0.25	2.88	No
4	0.01	0.42	0.05	1.68	No
5	0.30	0.97	1.50	3.88	Yes
6	0.81	0.93	4.05	3.72	No
7	0.49	0.09	2.45	0.36	Yes
8	0.40	0.64	2.00	2.56	Yes
9	0.35	0.81	1.75	3.24	Yes
10	0.72	0.02	3.60	0.08	No
11	0.80	0.14	4.00	0.56	Yes
12	0.00	0.99	0.00	3.96	No
13	0.10	0.07	0.50	0.28	No
14	0.51	0.04	2.55	0.16	Yes
15	0.82	0.94	4.10	3.76	No
16	0.69	0.91	3.45	3.64	Yes
17	0.64	0.15	3.20	0.60	Yes
18	0.90	0.46	4.50	1.84	Yes
19	0.10	0.43	0.50	1.72	Yes
20	0.10	0.41	0.50	1.64	Yes

The area of the rectangle surrounding the object is $4 \times 5 = 20$ square units, so the area suggested by this sample Monte Carlo is $20(12/20) = 12$ square units

11. Use the following scheme for generating card denominations

0000–0769 (A)	0770–1538 (2)	1539–2308 (3)	2309–2667 (4)
2668–3846 (5)	3847–4615 (6)	4616–5385 (7)	5386–6154 (8)
6155–6923 (9)	6924–7692 (10)	7693–8642 (J)	8463–9231 (Q)
9232–9999 (K)			

Using sets of five random numbers starting with column 3, one has 9771, 2783, 2752, 1218, 0526, which corresponds to a hand with K, 5, 5, 2, A. Twenty-five hands are generated with the 25th as 1170, 2404, 8531, 3736, 5973 or 2, 4, Q, 5, 8. Ten of these 25 hands consist of five different denominations of cards. The estimate is 0.400 when in reality it is about 0.507. Of course, with larger numbers of simulated hands the estimated probability should become closer to the actual probability. The goal here is a help to organize a Monte Carlo.

13. Let 0–199 represent 17 items daily. Then 200–349 represents 18 items, 350–499 represents 19 items, 500–749 represents 20 items, 750–899 represents 21, 900–949 represents 222, and a random number between 950 to 999 represents 23 items daily. The table summarizes the Monte Carlo and was generated using column 6 of the random number table as a start to simulate a month (30) of production.

RN	Items produced	Items moved	Items held over	Empty truck spaces
117	17	17	0	4
240	18	18	0	3
853	21	21	0	0
373	19	19	0	2
597	20	20	0	1
606	20	20	0	1
077	17	17	0	4
017	17	17	0	4
012	17	17	0	4
696	20	20	0	1
521	20	20	0	1
326	18	18	0	3
700	20	20	0	1
962	23	21	2	0
924	22	21	3	0
648	20	21	2	0
697	20	21	1	0
015	17	18	0	3
101	17	17	0	4
791	21	21	0	0
812	21	21	0	0
091	17	17	0	4
893	21	21	0	0
020	17	17	0	4
891	21	21	0	0
961	23	21	2	0
171	17	19	0	2
035	17	17	0	4
632	20	20	0	1
427	19	19	0	2

⟶

There were 10 held-over items over a 30-day period, so on average $10/30 = 1/3$ item awaits shipment per day. There were 53 empty spaces on the truck over the 30-day period or $53/30 = 1.77$ empty spaces per day on average.

15. Use the following scheme to convert the RN to a month of the year

000–083 (Jan)	084–166 (Feb)	167–249 (Mar)	250–333 (Apr)
334–416 (May)	417–499 (Jun)	500–583 (Jul)	584–666 (Aug)
667–749 (Sep)	750–833 (Oct)	834–916 (Nov)	917–999 (Dec)

Column 3 of the random number table is used to generate the groups of five people.

Trial	RN1	RN2	RN3	RN4	RN5	Hit or Miss
1	977 Dec	278 Apr	275 Apr	121 Feb	052 Jan	Hit
2	722 Sep	012 Jan	421 Jun	307 Apr	972 Dec	Miss
3	819 Oct	934 Dec	491 Jun	094 Feb	402 May	Miss
4	645 Aug	351 May	814 Oct	729 Sep	020 Jan	Miss
5	803 Oct	149 Feb	005 Jan	994 Dec	108 Feb	Hit
6	073 Jan	515 Jul	049 Jan	823 Oct	947 Dec	Hit
7	694 Sep	914 Nov	642 Aug	153 Feb	901 Nov	Hit
8	461 Jun	100 Feb	434 Jun	103 Feb	418 Jun	Hit
9	028 Jan	837 Nov	618 Aug	205 Mar	984 Dec	Miss
10	724 Sep	301 Apr	249 Mar	946 Dec	925 Dec	Hit

The simulation of ten groups of five people indicates a 6/10 chance that at least two share a common birth month.

17. If R_x is 000–499, the mouse makes a right turn and finds food. If R_x is 500–999, the mouse makes a wrong turn and continues through the maze until it finds food. Using column 1 to generate the necessary random numbers, one determines the following:

⟶

Trial	RNs	Right Turn?	Number of wrong turns
1	0.943 0.908 0.146	No No Yes	2
2	0.514 0.405	No Yes	1
3	0.733 0.044	No Yes	1
4	0.339	Yes	0
5	0.995 0.200	No Yes	1
6	0.477	Yes	0
7	0.212	Yes	0
8	0.094	Yes	0
9	0.361	Yes	0
10	0.634 0.137	No Yes	1
11	0.665 0.012	No Yes	1
12	0.665 0.012	No Yes	1
13	0.284	Yes	0
14	0.110	Yes	0
15	0.117	Yes	0
16	0.122	Yes	0
17	0.538 0.239	No Yes	1
18	0.089	Yes	0
19	0.458	Yes	0
20	0.240	Yes	0

There are nine wrong turns out of 20 trials for an average of 0.45 wrong turns per trial.

19. Answers vary.

EXERCISES 11.5

1. In the first stage, movement is from modes 2, 3, or 4 to node 5. The distance from 2 to 5 is 6, from 3 to 5 is 3, and 4 to 5 is 5. In the second stage, movement from node 1 to 2, 3, or 4 is possible. The distance from node 1 to 2 and then 5 is 14, from 1 to 3 and then 5 is 13, and from 1 to 4 and then 5 is 10. Therefore, the shortest distance is 10 if the path is from node 1 to 4 to 5.
3. There are two shortest paths at a distance of 8. They are 1–3–5–9–10 or 1–3–7–9–10 as shown in the following table:

Number of steps from node 10	*Distance to node 10*		LEAST COST TO NODE 10	
	From node	*Cost*	*Cost*	*Path*
	8	*4*	*4*	*8–10*
1	*9*	*2*	*2*	*9–10*
	5	$2 + 4 = 6$ $2 + 2 = 4$	*4*	*5–9–10*
2	*6*	$4 + 4 = 8$ $3 + 2 = 5$	*5*	*6–9–10*
	7	$1 + 4 = 5$ $2 + 2 = 4$	*4*	*7–9–10*
	2	$4 + 4 = 8$ $5 + 5 = 10$	*8*	*2–5–9–10*
3	*3*	$1 + 4 = 5$ $4 + 5 = 9$ $1 + 4 = 5$	*5*	*3–5–9–10* *or* *3–7–9–10*
	4	$3 + 5 = 6$ $2 + 4 = 9$	*6*	*4–7–9–10*
4	*1*	$2 + 8 = 10$ $3 + 5 = 8$ $6 + 6 = 12$	*8*	*1–3–5–9–10* *or* *1–3–7–9–10*

5. A total of \$6 million dollars can be spent on the plants, unused money is lost. Let $W(X)$ = Worth for the corporation for spending X dollars at Plant j at stage j.

   Y_1 = number of dollars allocated for stages 1, 2, and 3

   Y_2 = number of dollars allocated for stages 2 and 3

   Y_3 = number of dollars allocated for stage 3

⟶

$F_3(Y_3)$ is the optimal value for stage 3 given Y_3
$F_2(Y_2)$ is the optimal value for stages 2 and 3 given Y_2
$F_1(Y_1)$ is the optimal value for stages 1, 2, and 3 given Y_1

Working backward, one seeks $F_3(Y_3) = \max\{W(Y_3)\}$

Expenditure Y_3	Worth $W(Y_3)$	Optimal solution
0	0	$F_3(0) = 0$
1	5	$F_3(1) = 5$
2	5	$F_3(2) = 5$
3	7	$F_3(3) = 7$
4	7	$F_3(4) = 7$
5	7	$F_3(5) = 7$
6	7	$F_3(6) = 7$

Note that the only options that increase the worth are to spend one million or three million dollars. Spending two million for instance, is not enough for the three million dollar project so the total worth remains the same as for the one million dollar project.

Next one seeks $F_2(Y_2) = \max_{X \leq Y_2}\{W(X) + F_2(Y_2 - X)\}$ or the best value one can achieve using the various numbers of dollars and for a project at plants 2 and 3. The options are given as ordered pairs where (2, 1) represents that two million dollars is allocated to Plant 2 and 1 million dollars to Plant 3.

Weight Y_2	Worth	Optimal solution
0	$(0,0) = 0^*$	$F_2(0) = 0$
1	$(1,0) = 3 + F_3(0) = 3$ $(0,1) = 0 + F_3(1) = 5^*$	$F_2(1) = 5$
2	$(2,0) = 6 + F_3(0) = 6$ $(1,1) = 3 + F_3(1) = 8^*$ $(0,2) = 0 + F_3(2) = 5$	$F_2(2) = 8$
3	$(3,0) = 6 + F_3(0) = 6$ $(2,1) = 6 + F_3(1) = 11^*$ $(1,2) = 3 + F_3(2) = 8$ $(0,3) = 0 + F_3(3) = 7$	$F_2(3) = 11$
4	$(4,0) = 6 + F_3(0) = 6$ $(3,1) = 6 + F_3(1) = 11^*$ $(2,2) = 6 + F_3(2) = 11^*$ $(1,3) = 3 + F_3(3) = 10$ $(0,4) = 0 + F_3(4) = 7$	$F_2(4) = 11$

⟶

Weight Y_2	Worth	Optimal solution
5	$(5,0) = 6 + F_3(0) = 6$ $(4,1) = 6 + F_3(1) = 11$ $(3,2) = 6 + F_3(2) = 11$ $(2,3) = 6 + F_3(3) = 13^*$ $(1,4) = 3 + F_3(4) = 10$ $(0,5) = 0 + F_3(5) = 7$	$F_2(5) = 13$
6	$(6,0) = 6 + F_3(0) = 6$ $(5,1) = 6 + F_3(1) = 11$ $(4,2) = 6 + F_3(2) = 11$ $(3,3) = 6 + F_3(3) = 13^*$ $(2,4) = 6 + F_3(4) = 13^*$ $(1,5) = 3 + F_3(5) = 10$ $(0,6) = 0 + F_3(6) = 7$	$F_2(6) = 13$

Next one seeks $F_1(Y_1) = \max_{X \le Y_1}\{V(X) + F_1(Y_1 - X)\}$ or the best value one can achieve using the various sums of dollars for expansion projects at Plants 1, 2, and 3. The options are given as ordered pairs where (2, 1) represents two million dollars assigned to Plant 1 and the remaining dollars for expansion projects at Plants 2 and 3.

Weight Y_1	Worth	Optimal solution
0	$(0,\ 0) = 0^*$	$F_1(0) = 0$
1	$(1,0) = 4 + F_2(0) = 4$ $(0,1) = 0 + F_2(1) = 5^*$	$F_1(1) = 5$
2	$(2,0) = 5 + F_2(0) = 5$ $(1,1) = 4 + F_2(1) = 9^*$ $(0,2) = 0 + F_2(2) = 8$	$F_1(2) = 9$
3	$(3,0) = 7 + F_2(0) = 7$ $(2,1) = 5 + F_2(1) = 10$ $(1,2) = 4 + F_3(2) = 12^*$ $(0,3) = 0 + F_2(3) = 11$	$F_1(3) = 12$
4	$(4,0) = 7 + F_2(0) = 7$ $(3,1) = 7 + F_2(1) = 12$ $(2,2) = 5 + F_2(2) = 13$ $(1,3) = 4 + F_2(3) = 15^*$ $(0,4) = 0 + F_2(4) = 11$	$F_1(4) = 15$

→

Weight Y_1	Worth	Optimal solution
5	$(5,0) = 7 + F_2(0) = 7$ $(4,1) = 7 + F_2(1) = 12$ $(3,2) = 7 + F_2(2) = 15$ $(2,3) = 5 + F_2(3) = 16^*$ $(1,4) = 4 + F_3(4) = 15$ $(0,5) = 0 + F_2(5) = 13$	$F_1(5) = 16$
6	$(6,0) = 7 + F_2(0) = 7$ $(5,1) = 7 + F_2(1) = 12$ $(4,2) = 7 + F_2(2) = 15$ $(3,3) = 7 + F_2(3) = 18^*$ $(2,4) = 5 + F_2(4) = 16$ $(1,5) = 4 + F_2(5) = 17$ $(0,6) = 0 + F_2(6) = 13$	$F_1(6) = 18$

The best option for the Corporation is the \$3 million expansion at Plant 1, the \$2 million expansion at Plant 2, and the \$1 million expansion at Plant 3, which is worth of \$18 million.

7. Let $f_k(t)$ = min total probability of rejection when t hours of specialist time is allocated to job #k and subsequent jobs and $P_k(x)$ the probability of rejection when k hours are devoted to job k, $k = 1, 2, 3$. Starting with the time devoted to job #3 and assuming that it is the only job that is important in terms of rejection probabilities.

Time t_3	Probability of rejection	Optimal solution
0	0.55	$P_3(0) = 0.55$
1	0.50	$P_3(1) = 0.50$
2	0.30	$P_3(2) = 0.30$
3	0.20	$P_3(3) = 0.20$

At the next stage, it is assumed that the probability of rejection depends on job #2 as well as job #3. Ordered pairs represent the specialist hours spent on job #2 and the remaining on job # 3.

Time t_2	Probability of rejection	Optimal solution
0	$(0,0) = (0.45)(0.55) = 0.2475^*$	$P_2(0) = 0.2475$
1	$(1,0) = (0.35)(0.55) = 0.1925^*$ $(0,1) = (0.45)(0.50) = 0.2250$	$P_2(1) = 0.1925$

⟶

Time t_2	Probability of rejection	Optimal solution
2	$(2,0) = (0.25)(0.55) = 0.1375$ $(1,1) = (0.35)(0.50) = 0.1750$ $(0,2) = (0.45)(0.30) = 0.1350^*$	$P_2(2) = 0.1350$
3	$(3,0) = (0.10)(0.55) = 0.055^*$ $(2,1) = (0.25)(0.50) = 0.125$ $(1,2) = (0.35)(0.30) = 0.105$ $(0,3) = (0.45)(0.20) = 0.090$	$P_2(3) = 0.055$

At the last stage, it is assumed that the probability of rejection depends on job #1 as well as on job #2 and job #3. Ordered pairs represent the specialist hours spent on job #1 and the remaining time on the other two jobs.

Time t_1	Probability of rejection	Optimal solution
0	$(0,0) = (0.60)(0.2475) = 0.2475^*$	$P_1(0) = 0.2475$
1	$(1,0) = (0.40)(0.2475) = 0.099^*$ $(0,1) = (0.60)(0.1925) = 0.1155$	$P_1(1) = 0.099$
2	$(2,0) = (0.30)(0.2475) = 0.07425^*$ $(1,1) = (0.40)(0.1925) = 0.0770$ $(0,2) = (0.60)(0.1350) = 0.0810$	$P_1(2) = 0.07425$
3	$(3,0) = (0.15)(0.2475) = 0.037125$ $(2,1) = (0.30)(0.1925) = 0.05775$ $(1,2) = (0.40)(0.1350) = 0.0540$ $(0,3) = (0.60)(0.055) = 0.0330^*$	$P_1(3) = 0.033$

Note: the assumption that the final product is only rejected if all three finishing jobs are rejected.

The optimal solution is to assign no hours to job #1, 3 hours to job #2, and none to job #3.

Printed in the USA
K007834SCI081415 01S29053000000000645